W9-BYP-735

THE TEN TRUSTS

For Olivia
Follow your heart
Jane Goodall

THE
TEN TRUSTS

WHAT WE MUST DO
TO CARE FOR THE ANIMALS
WE LOVE

Jane Goodall and Marc Bekoff

HarperOne
A Division of HarperCollins Publishers

HarperOne

THE TEN TRUSTS. *What We Must Do to Care for the Animals We Love.* Copyright © 2002 by Jane Goodall and Marc Bekoff. All rights reserved. Printed in the United States of America. No part of this book may be used or reproduced in any manner whatsoever without written permission except in the case of brief quotations embodied in critical articles and reviews. For information address HarperCollins Publishers, 10 East 53rd Street, New York, NY 10022.

HarperCollins books may be purchased for educational, business, or sales promotional use. For information please write: Special Markets Department, HarperCollins Publishers, 10 East 53rd Street, New York, NY 10022.

HarperCollins Web site: http://www.harpercollins.com
HarperCollins®, ®, and HarperOne™ are trademarks of
HarperCollins Publishers.

FIRST HARPERCOLLINS PAPERBACK EDITION PUBLISHED IN 2003
Designed by Joseph Rutt

Library of Congress Cataloging-in-Publication Data
Goodall, Jane
The ten trusts : what we must do to care for the animals we love
Jane Goodall and Marc Bekoff.—1st ed.
p. cm.
ISBN: 978-0-06-055611-2
1. Human-animal relationships. 2. Animal rights. 3. Animal welfare.
4. Wildlife conservation. 5. Nature—Effect of human beings on.
I. Bekoff, Marc. II. Title.
QL85.G66 2002
333.95′416—dc21   2002068717

08 09 10 11 RRD(H) 10 9 8

This book is dedicated to the memory of those who taught me to respect and love the natural world: my mother, Vanne, who was my teacher, role model, and best friend for sixty-six years of my life; David Greybeard, Flo, and the other chimpanzees of Gombe; and that wonderful companion and teacher of my childhood, Rusty.

JANE GOODALL

I dedicate this book to my mother, Beatrice, who embraces all that is good about the world and who selflessly blessed my life with peace, warmth, compassion, respect, and bountiful love.

I am also deeply indebted to all of the extraordinary dog-beings with whom I have lived and the wild animals whom I came to know, awe-inspiring individuals who selflessly enriched my life by sharing with me the spirit, magic, and essence of who they are. And special thanks to Jethro . . .

MARC BEKOFF

CONTENTS

INTRODUCTION

This book, written in collaboration with my colleague and friend Marc Bekoff, is our joint attempt to argue for a closer connection with the natural world and a more ethical attitude toward all the creatures that make up the multitude of species with whom we humans share this planet. We shall try to show that teaching our children and all people compassion for other animals and respect for the places where they live will create a safer and more tolerant world. It is written for all who have hopes for the future, for their children and grandchildren. We believe that only when we understand can we care, and that only when we care sufficiently will we help. And we believe, too, that if we do not pull together to prevent the further deterioration of the fragile health of our beautiful planet, any hope for the future is in vain. We have reached a critical point in our history.

The book is based on a series of mantras that came to Marc's mind as he hiked with his canine companion, Jethro, in November 1999. Even as he enjoyed the mountain scenery of Boulder, Colorado, where he lives and teaches, he pondered the problems that plague those of us who care about the environment and the future of life on Earth. On the eve of a new year, a new century, a new millennium, Marc thought about all that was wrong in the world—the pollution, the destruction of habitat, global extinction of species throughout the world. For the first time in about three hundred years, in the year 2000, a primate had been pronounced officially extinct. Miss Waldron's colobus, a colorful monkey of the rain forests of Ghana, was gone—forever. People were shocked to a degree not experienced when they hear of the tens of thousands of less conspicuous species that have become extinct during the century.

Marc was dreaming of a time when scientists and nonscientists alike would work toward the same goal—creating a world in which people respect and live in harmony with the natural world, leaving lighter footsteps as they move through life. A world where the desperation of poverty and hunger is a thing of the past, and there is equitable distribution of those things necessary for a good life. Above all, a world in which we humans live in peace with each other, with animals, and with nature. The things he and I had so often talked and faxed each other about over the preceding few years. The mantras that obsessed Marc were a series of steps that we, the stewards of this planet, must actively take in our own lives in order to preserve and protect it. They described how we could all become better and more compassionate inhabitants of Earth. He asked me if I would add my voice—and the mantras that resulted were published at the turn of the century in the *Boulder Daily Camera*. This book was a natural next step, expanding the ideas of those mantras, which we came to think of as "trusts," as we are entrusted with the care of the animals and environments we love.

The Trusts' central theme is one I have been talking and writing about since I began to study chimpanzees in 1960—the importance and value of the individual. Not just the individual human being, but the individual animal being also. (Of course, we are animals too, but we use the word "animal" in its usual, everyday sense to refer to nonhuman animal beings.) And we argue that many animals, especially those with complex brains and central nervous systems, have person-alities, emotions, and the ability to solve those problems likely to crop up in their worlds. We humans have a more sophisticated brain when it comes to intellectual problem solving, but animals are not simply machines, marching to the tune of innate or inborn "instincts" or "drives"—they are able to make choices, to change direction according to need. Once we admit to this, we develop a new respect for these fel-low travelers and, with it, new ethical concerns about our so often abusive treatment of them.

When I was twenty-six, Louis Leakey sent me into the field to study chimpanzees. I had no scientific training and had not been to univer-sity. He wanted the observations of a naive mind, uncluttered by the reductionist thinking of the ethologists of the time. Initially, most sci-entists discredited my observations as anecdotal and unscientific, espe-cially when an article of mine appeared in the *National Geographic*. In those days, especially in Europe, it was bad form for a scientist to pub-lish anything in the popular press. I was written off as a "*Geographic* cover girl." But as my only goal was to discover the secrets of chim-panzee social life and I had already learned so much about animal behav-ior from my dog, Rusty, I was unconcerned by this attitude. The Ph.D. in ethology that I eventually earned from Cambridge University helped me to introduce my controversial ideas and methodology into main-stream science; compassion and science could work together after all.

Marc had a more conventional introduction to science than I did, but was also imbued with deep compassion toward the animals he studied. Gradually, over the years, he became increasingly disturbed by some of the methodology and the attitudes toward animals that he found in many of his colleagues in ethology and psychology. It was his collaboration with the philosopher Dale Jamieson that finally set Marc and me on the same platform—speaking out together for a more compassionate scientific attitude.

Marc and I have both met much criticism because we care about animals as individuals, because we believe it is possible to show empathy for the animals we study while maintaining scientific objectivity. We believe that the heart should be involved as well as the head. There is nothing unscientific in viewing animals as sentient and often sapient beings. It is an established fact that the physiological processes and the anatomy of the brain and central nervous system are remarkably similar in Old World monkeys, apes, and humans. There is a disturbing lack of logic in believing that, while the state of "clinical depression" may be similar in human and monkey, only a human is able to feel an emotion like despair or sadness. It is in fact more logical to assume that similarities in biology, brain, and the rest of the nervous system are likely to lead to similar repertoires of emotions in humans and animals—especially humans and apes. When an ape child behaves in the same manner as a human child, a way we describe as sad or happy, let the skeptics prove that the chimpanzee child is not feeling sad or happy.

One of our main goals in writing this book is to show that we humans are part of the wonderful animal kingdom and part of nature, not separate from them. We all share that mysterious elixir—life. We should like to pose a few theoretical questions; the way we answer these questions will be determined by the way we understand not only the place of animals in the world order, but ours too.

Most people, when asked whether they would risk their lives to rescue a drowning child, answer in the affirmative. Would you jump into icy sea, with waves pounding on the rocks, to rescue your child? Of course. But suppose it is a stranger's child? Or the child of your enemy? Or a child known to be in a vegetative state with no clinical hope of recovery? Most people insist they would risk their lives in all of these instances—although, unless we are actually faced with such a situation, we can never be sure. We do know, however, that we are expected to perform a rescue act of this sort.

But suppose it is your dog that is drowning, or some other dog? Would you jump in? Countless people risk—and sometimes lose—their lives in such a situation. People who risk their lives trying to rescue animals are not always praised. They may be described, disparagingly, as sentimental and foolish. After all, if they put their own lives at risk, then they, in turn, might need rescuing. They might put other human lives at risk. How do you feel about this?

Suppose your mother, your child, or some other loved one was dying and could be cured through animal experimentation—by sacrificing a chimpanzee or a dog. How much regret would you feel for the suffering of the experimental animal? Would you even think about it? Suppose your child was in a vegetative state and could be kept alive by sacrificing your dog—the one who had been your closest friend through long hours of nursing and agonizing over your child. If these were choices you could make within the law, what would you decide? Would your decision be influenced by what others would think of you? These are hard questions. But it is worth the mental exercise of trying to imagine how you would feel and act—and, more important, why. And how you would answer critics on either side of such a debate.

Dogs unhesitatingly risk their lives to rescue their human companions. Along with their incredibly brave handlers, the rescue dogs at Ground Zero worked courageously and tirelessly after the destruction of the World Trade Center. One, a nine-year-old Belgian Malinois named Servus, fell fifteen feet into a pile of dust. He was retrieved, partially suffocated. When he was eventually revived, his handler, Chris Christiensen, tried to take him home to rest. To his amazement, Servus refused to get into the car. "He just kept looking at me," said Chris. They ended up working seven more hours that day. Countless other dogs worked themselves to utter exhaustion and needed veterinary treatment before they could continue. One died.

Omar Eduardo Rivera was working at his desk on the seventy-first floor of the north tower when the hijacked airliner struck the building twenty-five floors above him. Mr. Rivera is blind, and his guide dog, Salty, was lying at his feet. Mr. Rivera said, "I stood up and could hear how pieces of glass were flying around and falling. I could feel the smoke filling up my lungs and the heat was just unbearable." And, of course, it was pitch dark in his sightless world. He knew he could not run down the stairs in the confusion of screams and shouting and rushing feet. He said he was resigned to dying, but hoped Salty might escape. "I unclipped his lead, ruffled his head, gave him a nudge, and ordered him to go." For several minutes Mr. Rivera struggled on his own in the chaos. Suddenly, he felt a familiar knee-high nudge. Salty had returned to guide his friend down into the street. "It was then I knew for certain," said Mr. Rivera, "that he loved me just as much as I loved him." The nightmare descent took an hour—they got out just before the building collapsed.

Dogs, along with many other animals, show no ambiguity when it comes to helping their human friends. But our actions toward animals are sometimes constrained by a fear of what others will think of us. Mr. Hoover, whose house was damaged and cordoned off after the fourth hijacked plane crashed in Pennsylvania, worried desperately about his cat, Woodie. He did not know if she was hurt, or even alive. Yet, in view of the human disaster, he felt it would be inappropriate to ask about a cat. But he confided in his parents, who asked for permission to go and search. This was refused, but a state police officer agreed to go to the house with a bag of cat food. Woodie survived, having been—unknown to Hoover—fed, watered, and petted by rescue workers overjoyed to nurture a life in the midst of so much death.

We humans in this modern Western world are confused in our thinking about animals. Most people love those animals with whom they share their homes, but sometimes seem unable to make the emotional and intellectual leap from their pet to other animals with whom they have no relationship. Thus a hunter can lavish affection on a deer fawn he has rescued, treat her as a part of the family, yet go out and shoot a deer of her species without remorse. A scientist who, behind the lab door, inflicts horrible suffering on a dog will come home and treat his own dog as a close companion, boasting about her intelligence, that "she understands every word I say." Hundreds of people watch the antics of birds on their bird tables, feed them through the winter, and provide nest boxes in the spring, yet never give a thought to the domestic hens, turkeys, and ducks who, in the nightmarish conditions of battery farms, live lives so cramped that they cannot spread their wings or roost or do any of the things that make avian life in the wild so joyous. Thousands of people who say they "love" animals sit down once or twice a day to enjoy the flesh of creatures who have been utterly deprived of everything that could make their lives worth living and who endured the awful suffering and the terror of the abattoirs—and the journey to get there—before finally leaving their miserable world, only too often after a painful death.

Once, when my ex-husband, Hugo van Lawick, and I were living and filming on the Serengeti, we came upon a Thompson's gazelle female trying to give birth. We watched her for more than two hours and then realized that the fawn was firmly stuck in a breech position. I pictured the darkness falling, the hyenas approaching, and the pathetic cries of the female as they bit into her living flesh. Of course, such things are truly part of nature and happen all the time. It is not helpful to adopt a romantic view of the lives of wild animals, for nature is, in many ways, "red in tooth and claw," even if, for the most part, we do not see the suffering.

Yet somehow Hugo and I felt personally involved in this particular drama. So we drove to see the park warden, hoping that he would end her suffering with his gun. But no—it was policy "not to interfere with nature." It would be wrong to deprive the hyenas of a meal. But, I argued, we would leave the body for the hyenas. Nothing moved him. Yet he had given permission to one scientist to shoot many wildebeests to examine their stomach contents at different times of year, and to another to capture vultures and inject them with the anthrax virus to see whether they could be artificially infected with a disease to which they appeared immune. A third researcher had permission to fly low over elephant herds, causing great distress, to mark them by dropping gallons of paint on their backs. The message was clear—it is okay to "interfere" in the name of science, in ways dictated by the head. But compassion, the messenger from the heart, must be allowed no voice.

But the world is changing. We are gradually becoming more aware of the damage we inflict on the natural world. This awareness is creeping into science, into the hearts and minds of the general public, and also into the legislation—all around the world more laws protecting animals and the environment are passed each year. For many species, though, our new understanding has come too late—they are gone. For thousands of suffering individual animals the pace of change has been too slow, but with the introduction of new measures we are gradually replacing cruelty with compassion and creating a world in which humans can live in peace and harmony with the natural world.

We know, Marc and I, that Rome wasn't built in a day. We realize, too, that change must come from within. Too often we see well-meaning and passionate animal rights advocates verbally abusing those they are trying to convert to their way of thinking. If we watch such an interaction, we see that very quickly those on the two sides of the argument have stopped listening to each other. Each side is too busy coming up with a new defense or attack. The exercise does little—except to harden the original attitudes of both.

It reminds me of an event that took place during my first visit to South Korea some five years ago. I attended a big press conference and was provided with a translator, a young Korean woman, who was very interested in animals. During the meeting I brought up the issue of the Korean practice of eating dog meat. The translator went quite pale. "I do not think you should talk about this," she said. "We are very sensitive about this." Well, of course I knew that. I assured her it would be okay, and she translated my message: "You know, in America and Europe people think that the Chinese and Koreans are very cruel because they eat dogs." There was a hush, and all the faces in front of me became tense and closed. "Well, in America and Europe people eat pigs. Pigs are every bit as intelligent as dogs. They can become friends with people, just as dogs can. I do not think it is ethically any worse to eat dogs than it is to eat pigs. If we eat animals at all, then surely the most important thing is how we treat them while they are alive, and how mercifully we kill them."

Not only did the Koreans relax, but they also began an animated discussion among themselves, which was passed on to me by my now relieved and smiling translator. First, some of the journalists wanted me to know that they too had dogs living in their houses. They did not eat *them*. One man said he had seen dogs outside a restaurant in very confined and bad conditions and that he felt this was wrong. Others agreed. There was a mention of some anticruelty law. I really had the feeling that this discussion might have sparked some new thinking. It was all based on the premise that people in glass houses shouldn't throw stones—or, as we say in Tanzania, "People in grass houses shouldn't light fires."

It is important to realize that many people perform cruel acts through ignorance—ignorance of the true nature of animal beings. Or because they do not think through the consequences of what they are doing. Or because they put up blinders, shielding themselves from the suffering that, if they will only admit it, they know they are causing or perpetuating. Like those working in medical research laboratories who believe that the suffering inflicted on animals is justified and try not to think about it. Sometimes, if people will only stop to think, there can be an amazing change of heart.

I was invited to participate in the Millennium Peace Summit of religious and spiritual leaders at the United Nations—over a thousand delegates from one hundred countries representing almost all the major religious and spiritual traditions. I was there to provide a voice for the "animal nations." I had only eight minutes to share my convictions that many animals have personalities, minds, and feelings—that their individual lives matter in the scheme of things. The following day, two African Catholic bishops approached me. They told me that when they were sick they sometimes consulted their traditional healers and were often told to sacrifice a chicken or a goat. But, in the light of my remarks, did I think it was right? At first I thought this was a question intended to lead to a discussion about humans having souls and thus being in a different category than animals. But they were genuinely interested in my thoughts. I told them that my role was to share my own belief concerning the true nature of animals and that the answer to their question was between them and God. They nodded and went off, seemingly satisfied.

A great deal of harm is being done to the environment, and thus to animals, in the name of "progress," for the economic benefit of those who are already wealthy and have more than they need, but continue to demand more. Meanwhile, the ever mushrooming human populations place increasing demands on dwindling natural resources. There are more than six billion of us now on the planet. In some areas, the environment is damaged by those who are desperately poor and struggling to make a living in areas where the land can no longer support their increased numbers. They cannot afford to buy food from elsewhere, so they cut down the trees to make space for their crops, houses, and livestock. Wild animals are hunted—shot, trapped, and snared—for food, but also by poachers anxious to make extra money by selling animals, or parts of animals, to anyone who will buy them. Unscrupulous and corrupt government leaders in some parts of the developing world pocket large amounts of money, much of which goes into private bank accounts, in exchange for leasing huge areas of virgin rain forest to foreign logging companies. Sometimes these companies are equally unscrupulous and practice clear-cutting, stripping the land and destroying habitats. As a result, deserts are spreading in the tropics and droughts and floods are becoming ever more frequent and more severe. Indeed, flooding as a result of ever increasing deforestation has become a huge annual problem all over the world, from Canada and mainland China to Bangladesh and Mozambique.

The accumulation of "greenhouse gasses," particularly from the burning of fossil fuels, is almost certainly contributing to the noticeable changes in global climate. This, in turn, affects a whole variety of animals. Let me give one example. Ian Sterling, a scientist with the Canadian Wildlife Service, found that rising spring air temperatures in the Eurasian Arctic and western Hudson Bay appear to be threatening the polar bear population. Polar bears hunt seals, waiting at gaps in the sea ice for them to come up to breathe and bask. When the ice breaks up, the bears can no longer hunt, and live mainly off stored fat until winter. The ice is breaking up two weeks earlier than it was just twenty years ago, the bears are coming ashore in progressively poorer condition, and their birth rate has decreased dramatically.

A recent survey by glaciologists, geologists, and other experts studying the ice in Antarctica concludes that there is a one-in-twenty likelihood of the West Antarctic ice sheet collapsing due to global warming—this would raise sea levels by about sixteen feet, with massive global consequences. I cannot forget the chilling words of Angaangag Lybetta, the leader of the Eskimo nation from Greenland.

In August 2000, he delivered this message to the gathering of a thousand religious and spiritual leaders in the United Nations General Assembly Hall: "In the north, we feel every day what you do down here. In the north the ice is melting. What will it take to melt the ice in the human heart?"

The Ten Trusts in this book are designed not just to "melt the ice" in our hearts. If we put them into practice in our lives, they will not only change our perspectives, but how we live on this planet. They are simple, yet profound. Taken together, they are about living with respect for *all* life, and they outline our responsibilities as stewards of the natural world. In the following pages Marc and I discuss many different aspects of our relationship with animals and nature. Although we describe a great deal of cruelty to animals, in all cases there is much we can do, as individuals, to change the status quo. As you read through this book, this will become clear. So much needs to change—sometimes just thinking about the ignorance and cruelty around the world can be depressing. So we have highlighted some of the many wonderful things that are going on and some of the inspirational people who are working together or separately for change.

We describe the Jane Goodall Institute's (JGI) Roots & Shoots program for youth, which now involves tens of thousands, from preschool through university, in activities designed to make the world a better place for animals, the human community, and the environment we share. We hope that all who read this book will realize how important it is for each and every one of us to do our part, so that we can move toward a world where cruelty and hatred are replaced by compassion and love.

Because change involves dialogue, we have written this book in our two voices. Mine will appear in regular type and Marc's in *italics*.

JANE GOODALL

THE FIRST TRUST

REJOICE THAT WE ARE PART OF THE ANIMAL KINGDOM

In our First Trust, we discuss some of the many similarities—biological, emotional, and intellectual—that clearly demonstrate the continuity of evolution not only in physical structure, but also in behavior. For those among us who do not believe in the theory of evolution, the anecdotes we have gathered together nevertheless provide compelling evidence of the many similarities in human and animal behavior. We can be proud that we are part of the animal kingdom.

Throughout recorded history the wise ones have known that we are a part of the animal kingdom. Native Americans and many other indigenous people of the world acknowledge their relationship with their brothers and sisters the four-footed ones and the winged ones and the finned ones. St. Francis of Assisi also described animals as his brothers and sisters, treating them with utmost tenderness and reverence. He rescued many creatures, from rabbits and lambs destined for slaughter to worms that were crossing the road. He empathized with them in sermons to the animals, birds, and fishes. The First Nation spiritual leader Chief Dan George urged his people, "If you talk to the animals, they will talk to you and you will know each other. If you do not talk to them, you will not know them. And what you do not know, you will fear. What one fears, one destroys."

Millions of people do not realize how closely connected we humans are with the rest of the animal kingdom. They do not realize that we ourselves are animals. Instead, they perceive a false reality in which humans stand on one side of an unbridgeable chasm and the rest of the animal kingdom stands on the other. Imagine a chimpanzee, so like us in so many ways, reaching out to us across that chasm. There is an unspoken question: "Will you acknowledge me as 'kin.'" If you dare look into his eyes and take hold of his hand, he will look back toward the other animal beings and then back to you with a question, "What about them? Don't they matter too?" Indeed, the great apes are like us in so many ways that they serve as ambassadors for all the other wonderful animals with whom we share the planet.

Of course, although we are animals, we are clearly unique ones. It is not just that we have a large and complex brain, but that somewhere in our evolutionary past we developed a sophisticated spoken language. Other animals with complex brains certainly have complex communication patterns, especially whales, dolphins, elephants, monkeys, and the great apes. Chimpanzees, bonobos, and gorillas even show the same lack of symmetry that humans do in a region of the brain (Broca's area) that is critical for speech production. In humans and these other three great apes, Broca's area is larger in the left cerebral hemisphere than it is in the right. But even these apes cannot, so far as we know, discuss the distant past, make joint plans for the distant future, teach their children about things or events that are not present, discuss an idea back and forth so that it can grow and change in accordance with the collective wisdom of the group, or ask why they are here. Nor, we suspect, do they worry about whether they have souls. Nevertheless, in so many basic ways our kinship with the world of animals, especially with the mammals, is striking. And nowhere is it more striking than with chimpanzees and the other great apes.

We share about 98.7 percent of our genes with chimpanzees, 97.7 with gorillas, and 96.4 with orangutans. We could have a blood transfusion from a chimpanzee if the blood types matched. They can catch or be infected experimentally with all of our contagious diseases. There are striking similarities in the structure of the brain and the central nervous system of apes and humans, and there are many similarities in social behavior and cognitive skills. Indeed, chimpanzees and the other great apes demonstrate many abilities that we used to think were unique to ourselves. They communicate by means of many different calls and also posture and gestures such as kissing, embracing, holding hands, tickling, swaggering, throwing, shaking a fist, punching, and so forth. They are capable of compassion and true altruism, but, like us, they also have a dark side to their nature and can exhibit brutality, and chimpanzees may even engage in a kind of primitive war. Although they have not developed a spoken language like ours (and cannot learn to speak words because of anatomical differences in the larynx), they have cognitive abilities that enable them to learn (in captivity) a variety of human languages, such as American Sign Language. They can make abstractions, generalize, and use abstract symbols in their communications. Some captive individuals enjoy drawing and painting.

Desmond Morris made a study of chimpanzee art back in the early 1960s and, as a joke, persuaded a London art gallery to hang one or two, by "An Unknown Artist." Critics, who spent a lot of time interpreting the meaning of this new art form, were somewhat embarrassed when they discovered the true identity of the unknown artists!

As a result of long-term studies on different populations of chimpanzees across Africa, we now know that there are differences in certain behaviors, such as tool using, that appear to be passed on from one generation to the next through observation and imitation—one definition of culture. Nor is cultural variation unique to chimpanzees. In a group of Japanese macaques living on Koshima Island, a young female, Imo, discovered that she could remove the sand from sweet potatoes by washing them in the sea. Other young ones imitated her, then their mothers and other young monkeys followed suit, and eventually the whole troop adopted this useful behavior. Subsequently, Imo learned to throw the corn that was scattered on the sand for the monkeys into the sea. The sand sank and she could pick up the clean corn. (Perhaps she also liked the salty taste.) This behavior was gradually incorporated into the repertoire of the whole group. Imo's behavior was first reported by Shunzo Kawamura in a paper published in 1954 in Japanese.

An amusing report of the spread of a newly learned behavior through a troop of rhesus macaques comes from the religious scholar David Haberman. He studied monkeys living in the temple town of Vrindaban, about a hundred miles south of Delhi. The monkeys, protected by the government, live in very large troops in which there is often fierce competition for food. In the early 1990s one monkey accidentally discovered that when he "stole" a pair of eyeglasses from a human, he was rewarded with food—bananas or roasted chickpeas—to return them. This happened again—and again. Then the behavior spread, and soon there were many monkeys in the troop who had learned this new way of making a living. The locals soon learned to remove their glasses. Then the monkeys had the wit to practice their new tradition on visitors.

Behavioral traditions unique to specific populations or groups are found in many animals other than primates. Female orcas, for example, spend years teaching their calves how to hunt elephant seals in accordance with the traditional methods of their group. Indeed, researchers have compiled a list of almost twenty behavior patterns in dolphins and whales that are influenced by local tradition and show cultural variation. Some female leopards have been taught by their mothers the dangerous art of killing and eating porcupines—in turn, they will pass the skill on to their cubs. Most leopards, though, would not dream of tackling such dangerous prey.

Chimpanzees are rather accomplished pharmacists. They practice self-medication, or zoopharmacognosy, by eating plants that cure them of stomach ailments. Michael Huffman, who works out of the Primate Research Institute at Kyoto University, once observed a wild female chimpanzee, Chausiku, who seemed to be sick. She slept while others fed. Later, as Chausiku traveled with her troop, she stopped and peeled the bark off of a mjonso tree, chewed on the pith, and swallowed the juice. The next day she was back to normal, eating ginger, figs, and grass. Huffman's local collaborator, Mohammed Kalunde, a national park game officer and herbal healer, told him that mjonso had medicinal qualities. Kalunde's people, the WaTongwe, use the plant, as do millions across Africa, to treat gastrointestinal disorders including malaria, parasitic infections, and upset stomachs. Huffman has also put forth his fascinating "Velcro theory" of leaf swallowing. He noted that many of the leaves chosen by chimpanzees have a bristly underside. He suggested that as the leaves pass through the intestine, they catch worms in the hairs and carry them along until they are excreted. In chimpanzee droppings Huffman discovered live worms entwined in the bristles of the leaves that had passed through the intestine.

ANIMAL EMOTIONS

Many animals in addition to humans show emotions that are similar, perhaps identical, to those we call fear, joy, happiness, embarrassment, resentment, jealousy, rage, anger, love, pleasure, compassion, respect, relief, disgust, sadness, despair, and grief. Indeed, it is the shared emotions, their expression, and similar physiological and anatomical bases that truly blur the borders between them and us.

It is no longer fruitful to ask if animals experience emotions, but, rather, why emotions have evolved—what functions they serve. The study of animal emotions, like the study of other behavior patterns, depends on a careful blending of anecdotes, common sense, and "hard" empirical data. None of these are dispensable, despite skeptics' denial of the importance of good animal tales.

FEAR

Fear is shown by all animals with complex brains, because "being afraid" is related to an individual's very survival. Often there are no second chances in nature—an animal has to perform the correct action the very first time he or she faces a dangerous situation (whether it is a predator or a stranger intent upon causing harm). Frightened animals will usually cringe, run away, or face the situation squarely. Some animals, such as opossums, just freeze in place to make themselves less obvious, hoping for the best. Orphan elephants who saw their mothers being killed often wake up screaming. Rescued street dogs also have nightmares and scream in their sleep.

It is often said that animals, including humans, emit strong odors when they are afraid and that other individuals smell fear. Jethro, my companion dog, a large German shepherd-rottweiler mix, enjoys going to the veterinarian, especially for acupuncture treatments for his sore elbow. But if the dog who had just been in the examination room was afraid, Jethro also shows hesitation and fear—he tucks his tail and pulls his ears back, rather than swinging his tail about knocking things off the examination table and holding his ears forward and alert. The smell of fear is conveyed by a glandular secretion produced by the anal gland of the previous canine client. Rats who have been exposed to cats show fear responses when they are exposed to the odor of a cat. Evolution—natural selection—has resulted in inborn reactions that are crucial to individual survival. There is little or no room for error when confronted with a dangerous stimulus.

THE JOYS OF PLAY

Social play is an activity that many animals enjoy immensely. Young chimpanzees and other animals tested in a laboratory will choose play over food unless they are really hungry. Play seems to be a self-rewarding behavior that is performed purely for the sake of engaging in it. Individuals become immersed in play, and there seems to be no goal other than to play. However, play is important for physical, social, neural, and cognitive development and may also prepare animals for unexpected situations. In Plato's Laws *he called for the formation of local play sanctuaries for the children of each village, areas in which to romp about and enjoy oneself that are reminiscent of modern-day dog parks for our canine friends. Play is that important for our and other animals' well-being.*

There is a feeling of incredible freedom in the flow of play. The activities look like fun and they surely are. Many of the same chemicals associated with joy in humans show increases in animals when they engage in play. Rats show an increase in dopamine activity when anticipating the opportunity to play. They also laugh when tickled.

Animals seek out play relentlessly, and when a potential partner doesn't respond to a play invitation, they often turn to another individual. If all potential partners refuse their invitation, individuals will play with objects or chase their own tails. The play mood is also contagious; just seeing animals playing can stimulate play in others. Animals seek out play because it is fun. Elk run across a snow field, jumping in the air and twisting their bodies while in flight, stop, catch their breath, and do it again and again. Bison, following one another, playfully run onto and slide across ice, excitedly bellowing "Gwaaa" as they do so. Chimpanzees may twirl around and turn somersaults when they play with friends or by themselves.

Many animals play only when they are young, but the great apes, dogs, whales, dolphins, parrots, crows, and ravens, among others, continue to play regularly as adults—just as we do. Adult female chimpanzees often play with their young, some more than others. Old Flo, at Gombe, sometimes found the play of her offspring absolutely irresistible. One day her family—adult Faben, adolescent Figan, and juvenile Fifi—began circling a tree near her, laughing and grabbing at each other's ankles. Flo was old, her teeth worn to the gums, yet she could not resist reaching out to grab the ankles of her romping family; after some minutes she got up and, with her infant, Flint, clinging to her belly, joined the game. My fondest memories of studying the hyenas at Ngorongoro Crater in Tanzania were the moonlit nights when the great females, after a successful hunt, returned to the den, where all the cubs from the neighborhood had gathered to play. And there the dominant females joined in the chasing games of their young, their huge bellies brushing the silvered grass.

When police are faced with a tense situation or are about to make an arrest, they will sometimes use humor to diffuse the tension, make a joke so there can be laughter. Animals sometimes do the same. Black Knight, a chimpanzee living in a zoo in Florida, regularly showed playful behavior when the alpha, or dominant, male was working himself up to perform an aggressive display. Such displays often ended with a brief attack. But Black Knight, noticing the bristling hair, furious scowl, and swaggering gait of his superior, would hurry over to stand upright in front of him, swaying from side to side and laughing. This ploy often worked, and what began as aggressive behavior ended in play.

EMBARRASSMENT: I DIDN'T DO THAT, DID I?

Can an animal really feel embarrassed? We feel embarrassment when we do or say something we wish desperately that we had not or when someone says something to us that makes us feel foolish. Animals do not blush from embarrassment, but certainly there are occasions when they do something that makes them undignified and then seem to hope they weren't seen doing it.

I remember one incident involving Gombe chimpanzee Fifi's oldest child, Freud. When he was five and a half years old, Fifi's brother Figan was alpha male of the community and Freud hero-worshiped this powerful uncle of his. On this occasion, as Fifi groomed Figan during a midday rest period, Freud climbed up the thin stem of a wild plantain. When he reached the leafy crown he began swaying wildly back and forth, just as Figan did during his impressive treetop displays. Had he been a human child, we would have said he was showing off. Suddenly the stem broke and Freud fell, landing close to where I sat. He was not hurt, but as his head emerged from the long grass he was looking toward Figan. Had his hero noticed? If he had, he paid no attention, but just went on grooming. Freud very quietly climbed another tree and began to feed.

Marc Hauser, of Harvard University, observed what could be called embarrassment in a male rhesus monkey. After mating with a female, the male strutted away and accidentally fell into a ditch. He stood up and quickly looked around. After sensing that no other monkeys saw him tumble, he marched off, back high, head and tail up as if nothing had happened.

When I study play, I analyze videotapes one frame at a time so I can see the details of the incredibly rapid exchanges that characterize these frolicsome encounters. Play bows are followed by a myriad of actions including biting, hip-slamming, mouth wrestling, faking left and going right, and exchanges of subtle facial expressions and eye movements. I am amazed at how many times I can look at a specific sequence and later see something I had not seen before. In one play sequence that I'd viewed about twenty times between Sasha, a rather large (some might say obese) malamute, and her friend Woody, a lithe, svelte mutt who visited her daily, I noticed that Sasha winced after Woody bit her on the lip. She continued to play for a moment and then turned away and licked her wound, tail still wildly wagging and her play face still in place. Sasha constantly looked over her shoulder as if to see if Woody were looking at her. And when he circled around in front of her, she turned her head to the side, licked once more, and then pounced on him as if nothing had happened. Wrestling play ensued, but no more biting. After they stopped playing, Sasha lay down, scratched her ear, and went to sleep. Later that day after Woody went back home, I saw Sasha lick her lip and wince. When I approached her, I noticed that she had sustained a deep cut on her lip, but there had been no way she was going to let Woody know that. Sasha had had enough and did not want to play anymore that day, even when her buddy Khartoum came up the road for his daily workout.

ANGER AND IRRITATION

Adult chimpanzees can be amazingly tolerant of youth, but sometimes get irritated. Once, as I was watching two young adolescents feeding, Evered, who was slightly older and stronger, suddenly attacked Pooch and snatched her bananas. Pooch screamed loudly, but soon things calmed down and they sat feeding quite close to each other. Suddenly, to my amazement, Pooch began screaming furiously, rushed at Evered, and started to hit him wildly. I looked up and saw the old male, Huxley, standing and watching. Huxley had become young Pooch's guardian after her mother died. Clearly, when Pooch saw him coming, she decided to try to enlist his help in punishing Evered. Huxley, however, sized up the situation, ran over and hit both the screaming adolescents—seemingly impartially—and peace was restored.

On another occasion, Fifi's son Freud, when he was just over five years old, climbed above a very large male baboon who was sitting on the ground. Freud, dangling from a low branch, began teasing him and kicking at his head. After a while, the baboon got annoyed, stood upright, roar-barked, and hit the irritating infant. Freud screamed, so Fifi, sitting nearby with her newborn, rushed over to defend him, barking and hitting at the baboon. Soon everything calmed down. Freud sat by his mother for a few minutes, then once more climbed to tease the baboon, and the whole sequence of events was repeated. I could hardly believe it when Freud, yet again, started his teasing. But this time, when the baboon lunged at Freud, Fifi hurried over and angrily hit Freud. He screamed very loudly, ran off, and eventually, thoroughly chastened, began to feed.

LOVE, SADNESS, AND GRIEF

Close and enduring bonds that can be described as love are found in many animals. Bernd Würsig observed courtship in southern right whales off Peninsula Valdis, Argentina. While courting, Aphro (female) and Butch (male) continuously touched flippers, began a slow caressing motion with them, and rolled toward each other. They briefly locked both sets of flippers, as in a hug, and then rolled back up, lying side by side. They then swam off side by side, touching, surfacing, and diving in unison. Würsig followed Butch and Aphro for about an hour, during which they continued their tight travel. Würsig believes that Aphro and Butch became powerfully attracted to each other and had at least a feeling of "afterglow" as they swam off. He asks, "Could this not be leviathan love?"

Love also abounds in relationships between mothers and their youngsters. In Thailand, a twenty-three-year-old elephant mother, Pang Soi Thong, became stuck after rescuing her calf, Lamyai, from the mud. Pang Soi Thong had broken free from her chain tether to rescue her daughter, who was mired in mud and screaming. When she herself got stuck, Lamyai helped to rescue her mother by pulling her to safety along with local villagers. The veterinarian who was at the scene, Alongkorn Mahannop, exclaimed, "The mother is healthy. She is safe because of the power of love."

Birds may also "fall in love." Konrad Lorenz, who won the Nobel prize for his work on animal behavior, observed that "the graylag goose's peculiar process of falling in love in many ways resembles its human counterpart." After bonding, males and females are strongly devoted to one another. As are many other species of birds, Bernd Heinrich writes of ravens: "Since ravens have long-term mates, I suspect that they fall in love like us, simply because some internal reward is required to maintain a long-term pair bond." Heinrich has studied and lived with ravens for many years and knows these wonderful birds well. Raven parents have to cooperate to capture prey for their young. They remain near one another during the day and also sleep next to one another and make soft vocalizations. They also play with and preen one another and share food. During courtship feeding they gently hold each other's bill.

In Tezpur, India, a troop of about a hundred rhesus monkeys brought traffic to a halt after a baby monkey was hit by a car. The monkeys encircled the injured infant, whose legs were crushed and who lay in the road unable to move, blocking all traffic. A government official reported that the monkeys were angry, and a local shopkeeper said, "It was very emotional. . . . Some of them massaged its legs. Finally, they left the scene carrying the injured baby with them."

Many animals become depressed and withdrawn when they lose a close companion. The intensity of the depression or grief depends on the strength of the bond between the two or the degree to which one is dependent on the other. The bond between chimpanzee mothers and their offspring is similar to that between human mothers and their children—in other words, it is similar to what we call love. Young chimpanzees who lose their mothers show many of the symptoms of clinical depression exhibited by human orphans: hunched and huddled posture, rocking, and withdrawing from social contact with their peers.

I shall never forget the days following the death of the old matriarch Flo, as we watched her son, Flint, sink deeper and deeper into depression and grief. A most poignant incident took place three days after her death. I watched Flint climb slowly into a tall tree near the place where she had died. Slowly, he walked along one of the branches, then stood looking down at an empty nest. After a minute or so he turned away and very slowly climbed to the ground. There he lay, his eyes wide and blank, their light dimmed. That nest was one he had shared with his mother a few days before she died. As the days passed, Flint became increasingly lethargic. He stopped feeding, showed no interest in the other chimps, and, with his immune system weakened, fell sick. The last time I saw him alive he was gaunt and hollow-eyed, sunk deep in depression and sickness. His last short journey was to the very place where Flo had died, by the clear waters of the Kakombe Stream. He stayed there for several hours, not moving, staring into the water. Then he struggled on a little farther, sank down onto the ground, and never moved again. I believe he died of grief.

The intensity of the love bond between whales is such that we would expect a corresponding intensity of grief when the bond is broken. Brenda Peterson describes a video, shot by a respected whale researcher, that captured an extraordinarily moving interaction between two humpback whales. The research vessel came upon the body of a dead male floating in the water. As the scientist prepared to examine him to look for the cause of death, a second humpback swam underneath the body and nudged it to the surface. Again and again the living whale repeated this tactic—it is what a mother whale does as she helps her calf to take its first breath. Then, for a while, the living humpback floated, motionless, beneath the dead body before rising up and embracing it with his huge pectoral fins. For five hours the great humpback floated there, fully embracing his dead companion. The researcher called the video "No Greater Love."

And, just as they love, so too can birds show grief for the death of a mate. Lorenz wrote: "A graylag goose that has lost its partner shows all the symptoms that John Bowlby has described in young human children in his famous book Infant Grief—*the eyes sink deep into their sockets, and the individual has an overall drooping expression, literally letting the head hang." Humans often exploit the loyalty between pairs of geese and ducks: after shooting one of a pair a hunter will wait near the dead body, confident the other will return. This is true, too, for other animals. Rick Bass, in his book* The New Wolves, *describes how the famous naturalist Ernest Thompson Seton once exploited the deep grief of Lobo, a bereaved male wolf. The body of Lobo's mate, Blanca, was dragged across the trap lines, spreading her scent. When Lobo returned, looking for his much-loved Blanca, he was trapped, tortured, and subsequently died. Seton and some of his friends had actually killed Blanca earlier. They had lassoed her and then "killed her by straining their horses in opposite directions 'until the blood burst from her mouth.'" Seton and his friends exulted in their slaughter of Blanca before they tortured Lobo.*

There are countless stories of the desolated grieving of dogs and cats when they lose a much-loved animal companion. Indeed, the American Society for the Prevention of Cruelty to Animals (ASPCA), in response to concern shown by humans about the deep grief that their dogs and cats express when they lose an animal companion, conducted a study to determine behavioral changes accompanying bereavement. Grieving dogs and cats show marked shifts in feeding and sleeping patterns; two cats starved themselves until they were so emaciated that they had to be euthanized despite all attempts to get them to eat. Grieving dogs and cats also showed changes in confidence and how much they solicited affection after their loss. For both dogs and cats, more than one-third of the animals demanded more affection when grieving. Fortunately, most behavioral changes in grieving animals due to the loss of animal friends are resolved in one to six months. But the intensity of their grief cannot be denied.

The devotion, loyalty, and love that can develop between humans and dogs is legendary. All of us who have been fortunate enough to share our lives with dogs and win their love are only too familiar with the terrible apprehension that comes over them when we start packing. My own Wiski-Biski would sleep on my suitcase, packed and ready in the hall. Once I forgot to zip it up, and he dug out all my clothes and curled up inside it. For a couple of days after I left, he always refused his food and his walks. But I think he always knew, deep inside, that I would return. And his welcome, when I did, was almost overwhelming. It was a demonstration of absolute joy.

Perhaps one of the most famous stories of dog grief is that of Greyfriars Bobby. John (Jock) Gray was a policeman in Edinburgh, Scotland, in the 1850s. Bobby was a Skye terrier. John died in 1858 of tuberculosis and was buried in Greyfriars Kirk graveyard. After the funeral Bobby went to the grave, and there he remained for fourteen years despite attempts to make him leave. Bobby even visited the restaurant where he and Mr. Gray ate lunch and, after feeding, returned to the grave. Bobby died in 1872 and was buried near his master. A year later a statue and fountain were erected near the entrance to the graveyard.

Michelle Rivera, author of Hospice Hounds, *relates the story of Sabrina, a nine-year-old Doberman pinscher, who became terribly depressed when her guardian of nine years passed away. The symptoms appeared immediately. Sabrina lay listless in her crate, eyes closed. Efforts to feed her became heroic as one shelter employee after another tried different methods of inducing her to eat. Sabrina was given fluids and a highly palatable, high-calorie, multinutritional food, but she did not react at all. She lay in her cage day after day, as the worried doctors and veterinary nurses debated her fate. Finally she was taken by the president of the local Doberman Rescue Concern, Nancy Armstrong. Sabrina finally recovered after being given special care by Nancy. Restored and renewed after her deep bereavement, Sabrina lived for three more years with Nancy, until she died of natural causes.*

Pepsi, a miniature schnauzer, was given by veterinarian Marty Becker to his father. Marty had helped to deliver Pepsi and knew him will. Pepsi had been the runt of his litter. Marty had to breathe life into him, as Pepsi was not breathing when he was born. Pepsi became his father's best friend. They shared the same food, the same chair, and the same bed. Marty's father died just after he turned eighty years old, by his own hand. Soon after family, friends, and the police left his house, Pepsi ran downstairs to the spot where Marty's father had died and stood as rigid as a statue. Marty picked Pepsi up, and Pepsi went from rigid to limp in his arms and emitted a painful moan. Marty put him in his father's bed, and Pepsi immediately went to sleep. Marty later found out from his mother that Pepsi had not been in the basement for ten years, for he was afraid of steps. He wondered what the connection they shared was. Had Pepsi gone to say good-bye? Pepsi never recovered from his companion's death. He died slowly, remaining weak and withdrawn. When Marty buried Pepsi, it all became clear to him—Pepsi had indeed died of a broken heart when the human to whom he was so closely bonded and devoted was no longer around to be his friend.

UNLIKELY FRIENDSHIPS

There are a number of examples of close and enduring relationships between members of different species in addition to those between animals and ourselves. It is common to find friendships between domestic animals living together in the same house, barn, or field. I knew one bitch who actually began to lactate when a tiny orphan kitten was introduced; their relationship as adults was a delight, for they played together like two dogs. James Herriot describes an enchanting relationship that grew up between a bedraggled kitten and a pig. And then there was the feral bitch who traveled for several weeks with a troop of vervet monkeys, with a small infant monkey clinging to her belly.

Among unlikely friendships, the strangest of all, surely, is that between Chino and Falstaff, a dog and a fish. Chino, a nine-year-old golden retriever who lives with Mary and Dan Heath in Medford, Oregon, and Falstaff, a fifteen-inch koi, have had regular meetings for the past six years at the edge of the pond where Falstaff lives. Each day, when Chino arrives, Falstaff swims to the surface, greets him, and nibbles on Chino's paws. Falstaff does this repeatedly as Chino stares down with a fascinated look on his face. Their close friendship is extraordinary and charming. When the Heaths moved, they built a new fishpond so that Falstaff could join them at their new home.

Dave Siddon owns and runs Wildlife Images, a wildlife rehabilitation center in Grants Pass, Oregon. In 1995, someone dumped four starving kittens at the center. Three were captured, but the fourth, Cat, eluded them. A few days later Dave watched in horror as the starving kitten squeezed through the fence and into the enclosure of a five-year-old orphaned grizzly bear, Griz, who had come from Montana after his mother and sister were hit and killed by a train. Cat went right up to Griz, who was eating his dinner. Then, in front of Dave's astonished eyes, the 560-pound Griz "pulled a little piece of chicken out and dropped it beside his forepaw, and the cat walked up and ate it." After that, these two orphans became close friends. They eat, sleep, and play together. Cat will not let humans near him unless Griz is there to protect him.

In the December 1994 issue of National Geographic *magazine, there is a delightful story and wonderful photographs of a thousand-pound polar bear playing with a Canadian Eskimo dog in Churchill, Canada. The dog, Hudson, approached the bear, wagged his tail, and grinned. He did a play bow, inviting the bear to play. The bear responded, and wrestling play ensued for several minutes. When it was all over Hudson and the bear embraced, after which the overheated bear lay down. Play between the bear and one of the dogs continued every evening for more than a week. When the ice formed, the bear was* able to leave for his winter hunting grounds. No one knows for sure why the bear and the dogs played, but they clearly enjoyed the games.

Clearly, mutually beneficial partnerships exist between animals. In the long run, what goes around comes around and we receive what we give. If we give love, we receive love, and in this cycle of giving and receiving more love is constantly being generated to be shared among all beings.

It is easiest to share these feelings of close connection with other animals when we recognize and appreciate that we are part of the animal kingdom. We can only tolerate so much alienation from our kin before the cycle of alienation reverses itself and reconnection and reconciliation are sought. Perhaps we humans view animals as having qualities we have lost—we yearn for their presence, their pure emotions, their zest for life. Let us rejoice that we are part of the animal kingdom.

THE SECOND TRUST

RESPECT ALL LIFE

In the Second Trust, we point out the value of every animal, plant, and tree in the complex web of life. Each living creature is worthy of our respect.

Every living being has its own spark of life. We humans look at the world around us and try to categorize, and thus simplify, the wondrous, diverse collection of life forms. We have large, highly developed, and complex brains with awesome capabilities for rational and abstract thinking. We have a highly sophisticated way of communicating with words. This has led, in the Western world, to a belief that humans stand in glorious isolation apart from the rest of the animal kingdom. We have placed the great apes on the next rung down, then monkeys, cetaceans, dogs, and so on until we get to insects, mollusks, and sponges. The worst part about this way of thinking is the conviction that we are superior to the rest of the animal kingdom. With our big brains and advanced technology, we have been able to dominate other life forms. And all those raised in the Judeo-Christian belief system have taken it for granted that Earth and all its riches, including animals, were created for our benefit. Western science gradually distanced itself from religion, especially in light of Charles Darwin's theory of evolution—but the belief in humankind's superiority remained.

Our domination of other animals is all but absolute. Not only can we kill them in the wild and destroy their homes, but also we can exert control over even the greatest of them. We know how to use pain to subdue and control. There are bulls, pigs, and camels with rings through their noses. There are cattle prods to send cruel stabs of pain through rebellious captive bodies. There are capture guns to anesthetize. There are whips and spurs to strike and goad. And, as a last resort, guns that kill.

We force the "beasts of burden" to carry us and our chattel or pull us over the ground—horses, donkeys, mules, camels, yaks, and elephants and dogs. Horses, dogs, and even dolphins have been forced to take part in terrible wars—and in the old days, as we know full well from Hannibal, elephants were often used in battle. Capuchin monkeys are taken from their mothers, have their teeth pulled, and are subjected to harsh training in order to become "helping hands" for paraplegics. Millions of animals are used in our research labs, from chimpanzees to rats, for the testing of drugs and vaccines, investigating disease, learning about the effects of space travel, and just finding out how they work. We dissect them to train our youth. We cage them in our zoos and stuff them for our museums. We train them, often with great cruelty, so that they can entertain us. And we breed them, in unspeakable captivity, to feed our ever mushrooming populations.

CALLING ANIMALS BY NAME

When I first got to Cambridge University in 1961 to work for a Ph.D. in ethology, I had no undergraduate degree. I had not been to college, and there were many things about animal behavior that I did not know. I had not been taught, for example, that it was wrong to give names to my study subjects—it would have been more scientific to give them numbers. I was dumbfounded by this practice. For one thing, I did not think of the chimps as "study subjects," but as individuals, each with his or her own personality. I was learning *from* them, not only *about* them. On a more practical level, I would never have been able to remember who was who if they only had numbers! Because we named them, many of the Gombe chimpanzees have become known to people around the world, thanks, largely, to the exposure in the National Geographic magazines and documentaries. A young woman approached me on the street during my first visit to South Korea and asked, "How is Fifi?" Somehow I cannot imagine her asking, "How is number 17?"

I was also reprimanded, at Cambridge, for ascribing personalities to the different chimpanzees—as though I had made up the vivid and unique characteristics of the various members of the Kasakela community! Only humans had personalities, I was told. Nor should I have been talking about the chimpanzee mind—only humans, said the scientists, were capable of rational thought. Talking of chimpanzee emotions was the very worst of my anthropomorphic sins. Fortunately, I had had a marvelous teacher in animal behavior throughout my childhood—my dog, Rusty. So I ignored the admonitions of Science. Today, though there are still pockets of resistance, most field biologists name their animals once they recognize them. There are studies of personality differences between individuals of the same species, and both the animal mind and animal emotions are now legitimate areas for scientific investigation.

SPEEDO THE CAT

When I first began conducting research in neurobiology and behavior, trying to figure out how cats processed visual information, I had no idea what I was

getting into. I'd teach cats to make discriminations among different visual pat-
terns, for which they were rewarded with food for making the correct choice.
Each cat had his or her own way of learning, some slowly, some rapidly, and
some not at all. I'd (secretly) name the individuals as they ran the maze, paying
attention to their personalities and learning abilities. I remember Speedo look-
ing at me when I lifted him from his small cage, anesthetized him, and then
proceeded to remove part of the visual cortex of his brain. As he succumbed to
the anesthesia, his eyes looked at me and asked, "What are you doing?" His
gaze is forever burned into my heart.

For a very short time I was able to continue this research—train a cat on a
particular task, remove part of the brain, and see how well they remembered
the task after recovering from the surgery. But it was when I had to euthanize
them (sacrifice—kill—with a minimum of pain, distress, and fear) to make
sure that the damage I caused was localized in the correct area of the brain that
it all came to a sudden halt. I did indeed regretfully euthanize four cats, Speedo
being the last. When I went to get Speedo for the final exit from his cage, his
fearlessness disappeared as if he knew that this was his last journey. His bold-
ness and cockiness melted as I picked him up, and tears came to my eyes. He
wouldn't break his piercing stare, and it broke my heart to kill him. I wish I
could have taken him home. To this day I remember his unwavering eyes—they
told the whole story of the pain and indignity he had endured.

So I left the program and entered another one in which naming was not only
permitted, but actively encouraged. I continued on various research programs,
including one studying Adélie penguins living in Antarctic, in which all of these
clowns were named, and others involving coyotes and various birds. Naming the
unique individuals I studied was as routine as eating breakfast before heading
out for a day in the field.

It was researchers' sudden passion in the 1960s for tramping into
wilderness areas to learn about the lives of animals in the wild that
started after World War II, that forced the scientific attitude toward
animals to change. Studies of chimpanzees, gorillas, orangutans,

wolves, coyotes, and whales disclosed that these creatures had amazingly complex societies and were, indeed, able to think and feel. The mechanistic, reductionist view that all animal behavior, although it might look human, was in fact dictated by simple stimulus and response was shown to be wrong. A new philosophy emerged as more and more scientists realized that animals with complex brains are, like us, able to experience pain and terror, contentment and joy—indeed, the entire emotional spectrum. Scientists began to admit that many species of animals do indeed have their own unique personalities and life histories.

Today science is beginning to "prove" what most people have always intuitively known: that we humans are not the only thinking, feeling beings on this planet. This leads to a new respect for the other amazing animals who are fellow travelers with us on life's journey. Charles Darwin himself said, "The love of all living creatures is the most noble attribute of man."

PORKY AND PHOENIX

Once an individual animal has been identified and named, there is an immediate change in the way he or she is perceived. This was illustrated dramatically during the terrible culling of animals enforced by the British government during the recent outbreak of foot-and-mouth disease. "Happiness As Porky Lives to Grunt Another Day" read the headline in one of our newspapers in late April 2001. Porky is a Vietnamese pot-bellied pig who lived with a group of pensioners. His doting friends kept him inside the house, and the men sent to kill him were barred from entering.

Porky was not alone; there was Phoenix, a white calf who was found still alive beside her dead mother five days after the herd had been slaughtered. Phoenix was headline news for days and the Ministry of Agriculture, responsible for the killing, was besieged with letters, begging—demanding—that Phoenix be spared. Even Prime Minister Tony Blair added his voice to the popular demand.

Public pressure saved the lives of Porky and Phoenix. Although millions of cattle and pigs were killed and people were horrified by the mounds of the dead, their hearts were not touched as they were by the plight of Porky and Phoenix—they were special because they were named and recognized as individual sentient beings. Most of the time the slaughter of animals for food goes on behind closed doors. We do not think of the individual beings who are being killed, often horribly, so that we can serve meat at our tables. We do not even talk about eating cows or pigs, but beef and veal, pork and bacon. That is why stories like those of Porky and Phoenix are important: they bring home the fact that animals are, indeed, individuals just like us, each with his or her own life filled, as ours, with happiness and sadness, fear and despair. And love.

ANIMALS AND ENTERTAINMENT

The general public has always loved to watch live animal shows. In the old days the theme was often man against beast. Gladiators fought fierce wild animals in the ring of the Roman circus. People thronged to watch Christians thrown to the lions. Bullfights are unbelievably popular in Spain. Rodeos, with their bucking broncos and other displays of cowboy skill, attract crowds in America. The animal acts at the circus have always been among the most popular—elephants standing on their heads, bears dancing, tigers jumping through flaming hoops, and the trainer putting his head in the lion's mouth. Many zoos around the world have animal shows, using penguins, parrots, birds of prey, dolphins, and many other species. Trained animals are also used extensively in movies and advertisements on television and in magazines.

There is often shocking cruelty in the way animals are trained to perform for the circus and other kinds of live animal shows. The cruelty takes place behind the scenes, particularly during pretraining. Incidents have been documented in which young chimpanzees were regularly beaten over the head with iron bars to instill a fear of the trainer that would lead to instant obedience during the show. A number of them died of a fractured skull. The orangutan Clyde, who appeared in the movie *Every Which Way You Can*, was unable to take part in promotional publicity. He apparently had been clubbed to death at the end of filming. The posed photos show his replacement, Dallas. There is also good evidence of serious abuse of chimpanzees during the filming of the movie *Project X*. Mary Chipperfield, daughter of the founder of Britain's oldest family circus, was actually prosecuted for her appalling abuse of a two-year-old chimpanzee who, understandably, did not want to go into her tiny, bleak, lonely, cold box each evening at four o'clock. At the same time, the Chipperfield elephant keeper was secretly filmed beating a chained elephant cow around the eyes and the base of her sensitive trunk with an iron bar. He beat her so forcefully the bar broke. Her crime? She had accidentally tipped over her water bucket. The keeper went to jail.

We, the general public, can make a big difference in all this. We need to begin by respecting all life. I remember when I spoke to eight hundred inner-city middle-school kids in south-central Los Angeles. To my surprise, many of them were angry to hear how chimps were trained for the circus. They understood why it was wrong to dress them in human clothes to make them look silly. I asked them what they could do to help. Many raised their hands: they would not go to see a circus with chimps; they would turn off the television; they would write letters. As a result of public concern expressed in these ways, a number of cities have banned circuses with exotic animal acts. India imposed a countrywide ban in 2000, and Rio de Janeiro in 2001.

Cirque du Soleil has developed a stupendous show that uses no animals whatsoever. The Chimp Channel (called "television's Missing Link") showed young chimpanzees dressed up and taught human mannerisms. It was on the air for only a short time as a result of the number of angry, concerned letters and e-mails that it elicited—many from our Roots & Shoots groups around the country.

In the United States sixteen Senators and fifty-five members of the House of Representatives demanded rapid federal action to rescue seven overheated polar bears from the Hermanos Suarez Circus, which was touring in Puerto Rico. The bears were kept in cramped quarters and in temperatures as high as 100 degrees Fahrenheit. Unfortunately, in early March 2002, a judge acquitted the circus owner of the charges leveled against him, and the bears still perform a ten-minute act, nine times a week. But, on the bright side, soon after this judgment, Alaska, one of the abused bears, was seized by U.S. wildlife officials and moved to the Baltimore Zoo.

More and more sanctuaries for ex-entertainment animals are being established. In Hohenwald, Tennessee, the Elephant Sanctuary provides a home for abused elephants. When Shirley, at age fifty-two, was taken to the sanctuary, she was put in a stall next to thirty-year-old Jenny. As soon as they met, they showed signs of extreme joyful emotion. With loud, rumbling roars deep in their bellies, they caressed each other with their trunks and remained in close contact, refusing to part. Indeed, Jenny tried to get into Shirley's stall. A search of records showed that the two had lived together in the same circus twenty-two years before, when Jenny was just eight years old. Given the proverbial elephant's memory, it should not be surprising that they remembered each other—it is the miracle of their reuniting that is so heartwarming. They remain close friends, rarely moving from each other's side.

Think of the mental suffering these elephants endured when, with no more thought than if they had been a pair of "things," they were separated and banished to different parts of the country. That, of course, is what we used to do to human slaves—and even today, in many parts of the developing world, thousands of children are still sold, especially in the sex industry.

Dolphins have big brains and the same kind of social needs as elephants. They too are captured from the freedom of the oceans and sent to endure lives of cramped and sterile misery in so-called oceanariums. "Water prison" is more like it.

Luna was a bottle-nosed dolphin. She and seven others of their kind were captured in Magdalena Bay, off the Pacific coast of Baja California, Mexico. The dolphins were transported in a truck to a small pen at the La Paz Dolphin Center to become tourist attractions. Luna arrived in a wooden crate, exhausted, frightened, and soaked in her own blood. She and her friends were kept in a narrow pen in water as shallow as forty-five centimeters. Luna died, and her death resulted in a moratorium in Mexico on the capture of dolphins in national waters. Project Luna was organized by Mexican environmentalist Yolanda Alaniz to close the La Paz center, and Mexican Environmental Minister Victor Lichtinger introduced emergency legislation to ensure that no other marine mammals shared the fate of Luna and her dolphin companions. In March 2002 the facility was closed.

The story of Keiko, star of Free Willy, *is too well known to repeat here. Although he still has not swum to freedom, the enormous effort made to rescue him from his tiny pool and reintroduce him into his native waters is heartening. We are not always cruel, and the more we understand, the more compassionate and loving we become.*

NATURAL HISTORY FILMS AND WILDLIFE PHOTOGRAPHY

Natural history films and photographs of wild animals are very popular. There is a band of professional (as well as amateur) natural history filmmakers and photographers whose work is truly astonishing, every film sequence, every shot, representing hours, weeks, months, and even years of patient observation in wilderness areas. I know some of them—Hugo van Lawick (to whom I was married), Michael Nichols, Michio Hoshino, and Tom Mangelsen. Their work is based on a love and knowledge of animals and their habitats. Unfortunately, though, there are others who show little respect for individual animals. One film team purchased a baby seal from an Inuit hunter when they wanted a scene showing a pup who had lost his mother. They filmed him on the ice and left him there to face a certain death. Another film crew attempted to show how seals and sharks swim in their underwater world by fitting them with small cameras known as "critter cams." Clearly, it is absurd to think that the highly stressed animals would behave normally. Several sharks actually died during the process.

Particularly shocking are the "game farms" that rent out animals for photography and filming, places where trained animals can be hired for closeup shots of "natural" behavior. One famous sequence showed a rented leopard "hunting" a baboon. To obtain the photos, which appeared in *Life* magazine, eight different baboons were used. They were literally thrown to the leopard, one by one, as the photographer clicked away until he got the shots he wanted. I do not remember how many of the baboons survived, but I do remember the extreme terror on the face of one of them—an old toothless male who was leaping high in the air as the leopard sprang at him. That photo won a big international award. In this time of increasing costs and competition, it is difficult to get funding for the extended time in the field necessary for a good documentary on animal behavior, and it is easy to see why captive and sometimes trained animals are rented. But it is unethical. First, the viewing public is deceived—for they believe they are viewing animals that are wild and free. Second, this material, obtained the easy way, compromises the careers of the genuine wildlife filmmakers and photographers. And third, the conditions in which the animals are kept are sometimes extremely bad.

Tom Mangelsen went to check out a game farm near Bozeman, Montana. It was even worse than he had imagined. Wolverines, wolves, lynx, coyotes, grizzlies, and other animals were pacing neurotically back and forth in cages standing under the hot sun. These cages were about 5 by 5 by 10 feet with iron bars and steel roofs and floor. Another photographer described how a mountain lion who had not performed properly was dragged through the woods with a chain around his neck—"to teach it a lesson." Another mountain lion died of lead poisoning—because his meals had included ground squirrels shot with a .22 mm rifle. Photographers and filmmakers typically want shots of young and healthy animals—what happens when they get old? Game farms of this sort become big business, saving money for the clients, making money for the owners—at the animals' expense. Pat Derby, president of the Performing Animal Welfare Society, who operates sanctuaries for retired and abandoned animals, believes that the "money is the bottom line" attitude makes it impossible to care for the animals properly. There is no respect.

Nonetheless, by and large, natural history documentaries have done a great deal to educate people about animal behavior. And many of these films also document the threats to animals and their environments and the often heroic efforts of the people who are trying to protect and conserve them. Most people do not have the opportunity or the desire to visit the really wild places of the world. National Geographic, Discovery, and PBS in America, Survival Anglia and BBC Wildlife in the United Kingdom, and many other channels in other countries bring the lives of wild animals into the living rooms of the world. Most film teams are comprised of truly enthusiastic and caring people who, to get their material, must struggle with heavy equipment in tough and sometimes dangerous environments.

SOME ASPECTS OF ZOO MANAGEMENT

In the United States, the American Zoo and Aquarium Association (AZA) is responsible for inspecting zoos, wildlife theme parks, and aquariums. If these institutions meet AZA standards, they are approved and accredited by the AZA. Unfortunately, there are about two hundred accredited zoos, "wildlife" theme parks, and aquariums in the United States, but nearly two thousand licensed zoos not accredited by the AZA. In Europe the same trend is evident— many more unaccredited zoos than the 280 accredited zoos.

Many unaccredited zoos are horrid places to live. However, even accredited zoos vary greatly in quality. Many zoo exhibits are antiquated and some zoo experts feel that only about one-third could generously be called "enriched" or "naturalistic." One zoo director says that he would like to be able to change 95 percent of the exhibits he has seen. Some zoos still get animals from the illegal animal trade, the "animal underworld." A 1995 Roper poll showed that about 70 percent of Americans were concerned about the well-being of animals in zoos. For good cause.

Animals in zoos surely suffer losses of liberty and privacy. Nonetheless, after Ryma, a much-loved giraffe, died at the Smithsonian Institution's National Zoo, the zoo's director said that she could not release Ryma's medical records to those who wanted to investigate the cause of her death because releasing them "would violate the animal's right to privacy and would be an intrusion into the zookeeper-animal relationship," a relationship similar to that of the confidential human patient-doctor relationship. This is an interesting and con-troversial tactic, one that was seriously questioned by many people. Courts of law do not recognize an animal's right to privacy. And if animals have a "right to privacy" then why are they placed on display without their consent for all to see as they eat, bathe, court, mate, and sleep? Could an animal ever agree to the release of his or her medical records? Could an animal's feelings be hurt if their records were released? If animals have a right to privacy then why don't they have the right to be free? These difficult and contentious questions deserve serious consideration for they raise important issues about animals' cognitive skills, emotional capacities, and moral and legal standing.

Although some zoos are working to preserve endangered animals, even animals in the better zoos are moved around like freight. For example, the Denver Zoo, in Denver, Colorado, had a series of very avoidable tragic events in spring 2001 resulting from social upheaval brought on by moving individual elephants in and out of their pachyderm facility. Mimi, the group's large matriarch, knocked over her friend, the elder and fragile Candy. Candy was so severely injured that she had to be euthanized (the zoo did not have the appropriate hoists for moving a large elephant). Mimi had expressed rage on a number of occasions after her herd mate, Dolly, was removed to go on her "honeymoon," the zoo's term for sending Dolly to another zoo to breed. In the meantime, Hope and Amigo, two new elephants, were moved in next door to Mimi. Mimi did not like it. No one paid attention, though. Sensing social unrest, Hope went on a rampage through the zoo in June 2001; mercifully, the six-thousand-pound mammoth did not seriously injure anyone, including herself.

Amigo had been taken from his mother when he was about two and a half years old. In the wild this separation wouldn't happen naturally. Elephant mothers are incredibly devoted, and Amigo should have been left at home with his own mother. In an interview I did with a local news station, I mentioned that Amigo had been ripped away from his mother. The spokeswoman for the zoo responded: "Amigo was never ripped away from his mother. He lives in the herd with his mom. He is here visiting the Denver Zoo for just three months." I wondered how Amigo could be with his mother and also be in Denver at the same time. On September 2 the zoo held a birthday party for Amigo, with a birthday cake and hat, and finally sent him back to his mother. The permissive regulations of the AZA still allow elephants and other animals to be moved around for profit.

Animals themselves are not always safe in zoos, but if they are not safe in zoos under human care, where can they be safe? In October 2001, also at the Denver Zoo, a male Asiatic black bear, Moktan, killed his long-time female cage mate, Sherpa. It was a horrific fight in which Sherpa sustained a crushed throat and mangled leg. Moktan and Sherpa had been left together despite a history of prior confrontations—thirty-six fights in the previous ten months. The zoo chose not to publicize the incident, claiming that they announce deaths only if the animals are "celebrities" like the elephants or polar bears. This unfortunate incident was discovered in the public records compiled by the United States Department of Agriculture (USDA), the agency that investigates such events. The Denver Zoo was fined a mere seven hundred dollars for its negligence.

FACTORY FARMING

We are a meat-eating culture. Based on evidence of stone tools uncovered in association with fossilized remains of prehistoric humans, it seems likely that we have hunted animals for food from the very beginning. There are still hunter-gatherer societies today whose lifestyles are probably similar to those of the earliest humans. They all hunt—and they all show respect for the creatures they kill. They offer up a prayer of thanks to the spirit of the animal who gave up its life that they might have food. In our modern culture we have lost that respect for each and every life. I was brought up to say grace before a meal, to give thanks for the food we were about to eat. But it was God we thanked. There was no sense of gratitude to the animals whose lives we had taken.

Throughout the developed world billions of animals are slaughtered for food annually. In the United States alone the numbers are horrifying: at least 93 million pigs, 37 million cattle, 24 million ducks, 2 million calves, 6 million horses, goats, and sheep, and nearly 10 billion chickens and turkeys are slaughtered for food each year. In 2001, more than 55,000 horses were slaughtered in the United States to meet consumer demands overseas. It has been estimated that more than 250 animals are slaughtered each second. The average American consumes about 2,400 animals in a lifetime. Agribusiness has become what Michael W. Fox, a senior scholar in bioethics at the Humane Society of the United States, calls a "slaughtering machine" committing violent acts each and every second of each and every day. What makes this all so fright-

ful is the conditions in which these animals are maintained—a direct result of our violation of the Second Trust.

Not only are billions of lives taken for food, but there also is a big problem with the enormous amount of waste material that accompanies meat production. Researchers at Colorado State University in Fort Collins, Colorado, estimated that a single 1000-pound beef cow produces 60 pounds of raw manure each day, about 11 tons per year. A dairy cow produces about 82 pounds per day, broiler chickens approximately 80 pounds of waste per 1000 pounds of chickens, and swine about 63 pounds a day. All totaled, meat animals in the United States produce upwards of 2 billion tons of waste each year, much of which filters into streams, lakes, and water supplies. This is about ten times the amount of waste that humans produce.

Beef cattle are often raised in "yards," where they mill about in small pens that are muddy and wet or baked dry and hard in the hot sun. Dairy cows are typically tied in stalls all day, every day, in long lines. Pigs, such intelligent creatures, are cramped in tiny, stinking pens as youngsters, then housed in cramped stalls or restrained under "farrowing hoops," that prevent sows from accidentally crushing a piglet (something they would never do under normal conditions) by preventing all movement. Calves destined for veal are raised in tiny twenty-four-inch-wide crates for the entire sixteen to eighteen weeks of their short lives. They can never walk, let alone frisk and gambol—they can't even lie down comfortably. They are fed a liquid diet twice a day. Prior to slaughter, iron intake is restricted to below normal levels and the calves become anemic, which results in the pale or white color of the meat; it is paleness that is the key factor in grading and pricing the carcass.

Poultry meat and eggs are now the most abundant and least expensive animal food products because of the development of a grotesquely huge industry devoted to their production. Hens raised for their eggs are kept in tiny, barren "battery" cages, which prevent them from expressing almost all normal behavior such as dust-bathing, perching, nesting—or even stretching their wings.

Often they have their beaks trimmed to reduce injuries and mortality associated with feather pecking and cannibalism. About half of the beak is removed using a hot cauterizing blade or a precision trimmer. The pain associated with beak trimming is thought to be intense and long-lasting. Caged birds often develop osteoporosis (weakened bones) because of a lack of exercise combined with calcium deficiency caused by their high rate of egg laying—they now lay upwards of 300 eggs per year, as compared to 170 in 1925. Some 25 percent of hens sustain broken bones when they are removed from their cages to be transported to the processing plant.

Animals are typically fattened up for human consumption using various hormones and antibiotics. Broilers now grow to market weight in about six weeks rather than sixteen weeks. A shocking incidence of little girls as young as five years of age in Mexico developing breasts has been associated with excessive use of hormones in the raising of chickens. Genetic engineering technology has also been used to produce animals who are bigger and more "meaty." Dairy cows stimulated with Bovine Growth Hormone (rBGR) can produce as much as 100 pounds of milk a day, about ten times more than they would normally yield. They suffer from frequent udder infections, and it has been estimated that almost all milk is contaminated with pus. When cows are treated with antibiotics these can be transferred to humans through the milk.

Factory-farmed food animals suffer physically and emotionally throughout their entire lives. And the slaughter at the end, before it brings the release of death, can be the worst torture of all. Often fully conscious hens and turkeys are shocked, scalded, and drowned in an electrified water bath as they are being prepared for market. Unwanted day-old male chicks are simply thrown into plastic bags and left to suffocate. Pigs and cattle are supposed to be stunned before being hung upside down by their hind legs in an assembly line leading to death. Gail Eisnitz, author of *Slaughterhouse: The Shocking Story of Greed, Neglect, and Inhumane Treatment Inside the U.S. Meat Indus-*

try, and the Humane Farming Association (HFA) of California have obtained dramatic evidence of shocking abuses in the IBP (originally Iowa Beef Processors) operation in Washington State. This evidence documents the illegal and tortuous mishandling, skinning, and dismembering of live cows in the strongest cruelty case ever assembled against a major U.S. slaughterhouse. Almost two dozen workers from the plant gave sworn affidavits, and hidden camera videotape documents the cruelty.

The conditions in the factory farms are so terrible that many animals die of disease or injury before they ever get to the slaughterhouse. Some 2.7 million calves die before they are one month old. It has been estimated that livestock losses due to disease total about $17 billion per year. That is the cost in dollars. We cannot measure the cost in animal pain and suffering.

Fortunately it is not only animal activists who are involved in trying to improve the miserable lives of animals who wind up in slaughterhouses. In a landmark speech delivered to the U.S. Senate in July 2001, Senator Robert Byrd (Democrat, West Virginia) railed against the rampant and barbaric institutionalized cruelty that it is inflicted on animals. Senator Byrd condemned the widespread violations of the Humane Slaughter Act and called for the USDA to end slaughterhouse cruelty. As chairman of the Senate Appropriations Committee, Senator Byrd asked for an additional $3 million for enforcement of the Animal Welfare Act and the Humane Slaughter Act. He also noted that animals suffer pain and made a plea for "respect for all life . . . and for humane treatment of all creatures."

SUNDANCE AND BUTCH

In January 1998, two pigs escaped from an abattoir in Wiltshire, United Kingdom, swam a river, and for eight days somehow eluded capture. The story captured the headlines and the hearts of the British people. As day followed day and the pair remained in the wild, there was a flurry of media attention. The pigs were named Sundance and Butch. Research into their history revealed that they were experimental pigs, crosses between Tamworth and wild boar, and had been escape artists from the start. On their eighth day of freedom the *Daily Mail* newspaper bought the two from their owner and sent out a special rescue squad. It was an all-night hunt in the pouring rain. Taking part were the police, a veterinarian, members from the Royal Society for the Prevention of Cruelty to Animals (RSPCA), a spaniel, a lurcher—and the media.

At the peak, 150 photographers and television crews were present from all the main British television channels and newspapers and some from Europe, America, and Japan. And, too, a huge crowd of local people. Butch, the female, was lured into captivity with food. Sundance, after breaking through the circle of well-wishers around him and hiding in dense undergrowth, was eventually flushed into the open and shot with a tranquilizer gun. Two days later the pair was sent to an animal sanctuary to live out their lives in peace. Quite apart from the drama, the story of Butch and Sundance started a wave of interest in vegetarianism. "One of the beautiful things about the British people is that it doesn't take much to prick their consciences," said a spokesman for the Vegetarian Society. "Almost the entire population was rooting for these pigs, and people are also making the connection to eating meat." That is pork, bacon, ham—pig.

WEARING FUR

Human beings have always worn animal skins for warmth or adornment. Indeed, it was only by wearing the fur of arctic animals that people such as the Inuit were able to invade the icy northern wastes. Unfortunately, wearing fur became fashionable in the wealthy Western world. A great many animals are trapped for their fur, suffering horrible deaths in leghold traps, wire snares that get pulled tighter and tighter around the neck or leg as the animal struggles, and conibears that rip the entire body and break the neck or back. Beavers are often trapped in water and eventually drown. In the United States there are no laws regarding the killing of trapped animals—they can be killed in any way the trapper pleases. Additional suffering is caused when domestic dogs and cats get caught accidentally in traps set for other animals.

Animals of many kinds, including dogs and cats, are bred on farms for their fur. Conditions on these farms are similar to those that we have described for factory farming of food animals. The fur industry has issued a set of guidelines for the treatment of farmed animals, but their use is voluntary, and fur farms are not monitored. The animals are killed in a variety of ways: they may be beaten over the head, suspended and hung from a noose, or bled to death. Sometimes they are gassed or given lethal injections. Mink are often killed by neck snapping. And just as cows and pigs panic as they are driven to slaughter, so, too, do fur-bearing animals like mink when they are removed from their cages; they screech, urinate, defecate, and struggle desperately to escape.

Animal rights activists have fought long and hard to create an awareness of the suffering involved in the making of fur coats. Some of their highly emotive advertisements included one in which a model's fur coat dripped blood behind her as she walked and one in which an actress at a fashion show shocked and delighted the audience by preferring to appear nude rather than wear fur. For a while people responded and fur sales dropped during the 1980s and 1990s. But as we move into the twenty-first century the Fur Information Council of America reports a revival of the industry—in 2000 sales of fur coats totaled $1.68 billion, up 20 percent from the year before. That means that in one year between 7 and 8 million animals were killed for fur in the United States alone. And worldwide we learn that

28 million animals were farmed, and 7.6 million trapped, for their fur. Yet we do not need to wear fur. The days when people had to wear animal skins in order to survive the cold are gone, except among the indigenous people of the far north. For the rest of us, synthetic fabrics can be just as effective.

If fashion designers, models, and those who buy fur were forced to watch an animal struggling in the jaws of a steel leghold trap or the often brutal killing and skinning of farmed animals, if they had to listen to their screams, I wonder how much difference it would make? How many would shudder and turn away, complain that they were sensitive and could not bear to think about such things—and continue to see, model, buy, and wear fur?

But there are signs of hope. Some seventy countries worldwide (but not the United States) have banned leghold traps and twenty have banned all forms of trapping. Fox farming is now banned in Holland, and Italy has banned the use and sale of dog and cat fur. Fur farming has been banned in the United Kingdom. And in New York City, some exclusive nightclubs do not allow people wearing fur to enter their establishments.

ANIMAL EXPERIMENTATION

In 1996, according to a survey by the USDA, about 1.3 million individuals, including 52,000 nonhuman primates (about 2000 chimpanzees), 82,000 dogs, 26,000 cats, 246,000 hamsters, and 339,000 rabbits were used in experimental research. These numbers don't include the tens of millions of rats, mice, and birds—about 95 percent of the animals used in experimental research—animals who are not protected during experimentation and will not be protected anytime soon. In February 2002, the U.S. Senate voted to continue to exclude rats, mice, and birds from protection under the federal Animal Welfare Act. Jesse Helms (Republican Senator from North Carolina, a state in which researchers use about 250,000 of these animals in research), who intro-duced the bill, claimed, "A rodent could do a lot worse than live out its life span in research facilities." And he said by excluding these animals from protection biomedical researchers wouldn't have to put up with "regulatory shenanigans" put forth by the "so-called animal rights crowd."

Nonetheless, it is important to note that in a recent poll conducted by the psychologists Scott Plous and Harold Herzog, a majority of researchers them-selves favored federal regulation of research conducted not only on nonhuman primates (99.7 percent favored regulation), dogs (98.6 percent), and cats (98.3 percent), but also of research performed on rats (73.9 percent) and birds (67.9 percent).

In the United States alone, it is estimated that more than 70 million animals are used annually in research, and that one animal dies every three seconds, often in pain and fear. There is inconsistency in the ways in which research pro-tocols are evaluated by Institutional Animal Care and Use Committees (IACUCs) in different universities in the United States. Some projects that are approved in one university are not permitted in others. In the United Kingdom, according to a report in the Sunday Independent *newspaper, more than 6.5 million mice, 2.4 million rats, and 1000 dogs were killed because they were not needed for research after being bred for such purposes. Many spent their entire lives waiting to be used before being deemed "useless" and killed.*

The U.S. military kills an average of five nonhuman primates each day in its research laboratories, and the numbers are increasing. About 1,500 individu-als were used in 1997 and 1,877 in 1999. Military experiments subject ani-mals to, among other things, chemical weapons, ionizing radiation, and lasers and high-power microwaves. Many of these procedures are extremely invasive and extremely painful, yet according to the USDA, more experiments are con-ducted without pain relief by the military than in other research laboratories—about 9 percent of the animals receive no pain medication. In the thirty-four laboratories run by the Department of Defense (DOD), about one-fifth of all animals are used in painful experiments without anesthesia. In 1999 more

than 327,000 animals were used by the DOD, a 12 percent increase over the
previous year. We can expect numbers to rise sharply in response to new fears
of biological and chemical warfare.

The United States is but one of the many countries that practice vivisection. Exact numbers are often hard to get because, in many countries, records are poorly kept. Some examples: in 1991 Japan used 12,236,000; France used 3,646,000; and Greece used 25,000. In 1994 Sweden used 352,000. It is good to know that many people are engaged in thousands of campaigns designed to improve conditions in the laboratories and, ultimately, to eliminate the use of all live animals. If only the pioneers of vivisection had recognized an animal's capacity for pain and suffering, this huge animal research industry (and it is a multibillion-dollar industry worldwide, with countless individuals making huge amounts of money) might never have developed in the way it has. The fact that it continues, despite our new understanding of animal nature, is a sad indictment against humanity.

Fortunately new alternatives to the use of live animals are being developed all the time. Proponents of animal experimentation usually say that, unfortunately, it will always be necessary to use some animals, but that efforts will be made to use as few as possible and treat them as humanely as possible. We need a new mind-set: let us admit that what we are doing is unethical and use our extraordinary brains to find ways to eliminate the practice of using live animals as quickly as possible. More money for developing alternatives and a few Nobel prizes for success would help.

Nonresearchers are becoming increasingly concerned with animal experimentation. A recent poll shows that 75 percent of people questioned in the United States disapprove of experiments that subject animals to severe pain and distress, and 60 percent oppose experiments that cause even moderate pain and distress. These results parallel polls taken in the United Kingdom. A growing number of scientists in the United States oppose research that causes pain and death—support among psychologists and psychology students declined almost 50 percent for research causing pain and death to animals.

GRANTING APES LEGAL RIGHTS

In 1993 The Great Ape Project: Equality Beyond Humanity *was published. This, in turn, launched a project known as the Great Ape Project (GAP). GAP is fighting to have the great apes admitted to the Community of Equals—urging that certain basic rights that are extended to humans and that are (theoretically) enforceable by law be extended also to the great apes: the right to life, the protection of individual liberty, and the prohibition of torture. In other words, the great apes should be given equal moral standing with humans and equal consideration under the law as to how individuals should be treated. Seven years later, Steven Wise, one of the growing numbers of animal rights lawyers, published* Rattling the Cage. *In this book he presents compelling evidence as to why chimpanzees should be granted legal standing— because of their rich social lives, their humanlike intellectual abilities and emotions, and their capacity for mental, as well as physical, suffering.*

What is the use, you may ask, of extending rights of any sort to great apes when there are so many horrifying abuses of human rights around the world? Why don't we simply urge people who care to become more responsible? I think the legal aspect is very helpful too. Many people are astounded when they hear that lawyers from respected law firms are actually working, pro bono, on questions pertaining to the rights of animals—there must be something in it after all! Thus, more and more people have been jolted into thinking in new ways. Maybe this played some part in getting support for two important bills that were passed in 2000 (by both the U.S. House and Senate), the Great Ape Conservation Act and the Chimpanzee Health Improvement, Maintenance, and Protection Act (the CHIMP Act).

The first provides for certain monies to be budgeted for conserving the great apes in the wild. The second concerns the U.S. government's commitment to provide sanctuaries into which great apes from biomedical research can be retired. Although there is still much work to be done to provide greater protection for chimpanzees who have been used in research, this is a much-needed beginning.

Why this concentration on the great apes? Don't other animals matter too? Of course they do, but great apes, being our closest living relatives and sharing with us many biological, cognitive, and emotional characteristics, provide the best starting point if we are trying to demonstrate that there is, after all, no sharp line dividing humans from the rest of the animal kingdom, at least from a scientific or logical perspective. From an emotional perspective, too, we need only to look into the eyes of a chimpanzee to know, intuitively, that we are looking into the mind of a thinking, feeling being. The more we learn about their behaviors, especially their qualities of compassion and altruism, the more compelled we are to try to protect them from human exploitation. Once the apes have helped us to understand that there is no clear-cut division between humans and the animal kingdom, we gain new respect for the other amazing creatures with whom we share the planet.

David Greybeard was the first chimpanzee who learned to trust the strange white ape who had invaded his world. After about a year he actually allowed me to follow him through the forest. One day, after struggling after him through a tangle of thorny vines, I found him sitting, almost as though waiting for me. Perhaps he was. As I sat near him, I saw the ripe red fruit of an oil nut palm lying on the ground. I picked it up and held it toward David on my palm. He turned his head away. I moved my hand closer. He turned, gazed directly into my eyes, took and dropped the nut, then very gently pressed my hand with his fingers. This is how chimpanzees reassure each other. His message was clear: he did not want the nut, but he understood I meant well. I had communicated with David in an ancient language that predated words, a language that we had inherited—the human and the chimpanzee—from some common ancestor who had walked Earth millions of year before. David Greybeard has long since gone to the Happy Hunting Grounds, but I have honored his trust through all the years after his death.

In early June of 1978, I walked among sagebrush around Blacktail Butte, just south of Jackson, Wyoming. I was looking for Sally, a coyote mother who'd just given birth to a litter of pups, as she had in two previous springs. I figured that if I found Sally, I'd find her den. I wanted to see how many pups there were and record data on social development, play, and maternal behavior. I found the den, and as I approached it, I felt the presence of another animal. I had not seen, heard, or smelled Sally or any other coyote, but I had this unnerving feeling that someone was there. Then I heard a soft bark. I turned around and saw Sally watching me from about ten meters away. She stood upright, ears forward, cocked her head from side to side, barked again, and stared directly at me, her brow furrowed with an unwavering, intense gaze. We connected intimately. I was shaken, not out of fear, but rather in awe of the strength of her presence. "What are you doing—get away from my kids," she seemed to say. Sally continued to stare, and our eyes locked for what seemed to be an interminable amount of time. I broke the mutual gaze, walked slowly away, and felt badly for intruding into Sally's inner sanctum. From that day on, my research crew and I stayed away from dens until the pups were out and about, and Sally paid us little or no attention in the future.

We are beginning to learn that each animal has a life and a place and role in this world. If we place compassion and care in the middle of all our dealings with the animal world and honor and respect their lives, our attitudes will change. All lives become sacred and we become more careful with our care of them.

THE THIRD TRUST

OPEN OUR MINDS, IN HUMILITY, TO ANIMALS AND LEARN FROM THEM

In the Third Trust, we show how animals can comfort and heal us as well as fill our lives with love and joy. If only we will accept them as companions and not slaves who must, at all costs, be dominated, we can learn so much from them and the stories of their lives.

A book from my childhood, *The Miracle of Life*, opened my eyes to wonders of nature, the tremendous diversity of life on Earth, and the endless adaptations of behavior and structure in the animal kingdom. Animals have adapted to live in extremes of heat and cold and extremes of altitude. We find them fathoms deep in the seas, high in the mountains, in the burning sands of the desert, and roaming the polar wastes.

Gradually, through evolution, some fins became hands, some hands then became leathery or feathery wings. Others became flippers, so that their owners could go back to life in the water. Some animals walk on a hundred legs, others run on two. Some animals have five toes, or four, or three, or two, or one toe on each foot, and others glide along with no visible legs at all. Animals sleep on the ground, in burrows, in holes in trees, in caves, in elaborate nests, under the water, even cradled in air currents.

Animals have developed an amazing variety of tools, weapons, and other structures that enable them to hunt, catch, prepare, and eat the various items of their incredibly diverse menus. I especially love the fish that spits drops of water at insects as they fly past, knocking them into the water. There is a spider who whirls a strand of silk with a blob of glue on the end around and around her head to snare a passing fly—like a little cowboy with a lasso. There are hunters who stun their prey with electric jolts, inflict poisonous bites, run like the wind, or wait patiently, camouflaged to look like their surroundings. And, of course, the prey animals have adapted in countless ways to escape becoming food for the hunters.

We could go on and on without coming to the end of the diversity of animals, of structures, of habitats. The longer I live and the more I learn, the more awed I become. What can be more extraordinary than the migrations of birds and fishes, and even insects? Young animals setting off to travel thousands of miles from their birthplace to traditional feeding grounds, finding their way with the help of stars and the magnetic field. Even delicate-seeming butterflies migrate—most of us know of the migration of the monarch butterflies who return to the same resting places every year even though each group is three generations removed from the previous visitors. And we are all familiar with some of the truly mind-boggling journeys that have been undertaken by dogs and cats to return to their old homes.

ANIMAL COMMUNICATION

I am especially fascinated by the numerous ways in which animals communicate with each other. These include a truly vast repertoire of vocalizations ranging from glorious singing to the most raucous of squawks, and, too, sounds made in other ways, such as the vibrating of membranes, the scraping of legs together, drumming on tree trunks, and so on. Then there are the postures and gestures, starting with those with which we are most familiar—the humanlike repertoires of the apes and monkeys, the wagging tails of dogs and cats (which must be interpreted so differently in the two species), the extraordinary posturing of some birds during courtship. Nor must we forget the olfactory communications, the production of scents secreted by various glands and left around to mark the individual's territory. Some female insects send out scent signals that can attract cruising males from miles away. Pheromones are chemical signals that indicate reproductive status. More exotic is the incredible waggle dance of the honeybee that indicates the distance of food sources from the hive and the precise direction the other bees should fly. Firefly females flash little lights, and some fish glow eerily with strange phosphorescence in the dark, cold world far down in the deep waters where until recently it was thought there could be no life.

The accomplishments of the animals are almost overwhelming. Indeed, many of our human innovations may actually have been learned from observing animals. Weaverbirds were undoubtedly fashioning their elaborate nests long before people began weaving baskets. The first human being to shape a pot of clay may well have been inspired by the painstaking labor and elegant shape of the potter wasp's nest or perhaps that of the house martin swallows or some other bird that builds its nest of mud.

It seems, too, that some animals have strange abilities of prediction. In China it has long been known that animals are able to predict earthquakes. Some dogs and cats can accurately predict the onset of a seizure in epileptics. And the scientist Rupert Sheldrake has investigated thousands of reports of animals—mostly dogs—who seem to know, within seconds, when their owners are starting out for home.

CONVERSATIONS WITH N'KISI

In January 2002 I met N'kisi, an African gray parrot, bred in captivity, who lives with Aimee Morgana and several other birds. I had heard that N'kisi carried on conversations and, most improbably, communicated telepathically with Aimee. I spent an hour and a half listening to Aimee talk about N'kisi, while he talked, mostly to himself, but sometimes directly to Aimee and to me. His cage—which he can enter and leave at will—stands up high, so that he can feel dominant over humans. A few times he flew over to join us. One of the first things N'kisi said as I was introduced to him was "Got a chimp." Aimee was delighted, because she had been showing N'kisi photos of me and the chimpanzees. "Chimp" was his 701st word, and this was the first time he said it! While I was there, I gave a chimpanzee greeting call, a pant hoot. N'kisi was fascinated. Six weeks after that visit, Aimee tells me, N'kisi is making that sound several times a day with comments like, "That's a chimp. Sounds like Jane." He tends to talk about special experiences for months afterwards.

Aimee keeps a detailed diary and shared some stories. She makes exquisite jewelry. One day a particularly lovely necklace she was wearing broke. As Aimee picked up the beads, she was thinking what a shame it was that it broke, as she just made it. N'kisi said with a sympathetic tone, "Oh no, you broke your new necklace." Another time, Aimee was showing N'kisi some little candles she had bought, and he said, "Look at the little candles." Then he said, "Look at the little bug." One of the candles had an image of a spider on it for Halloween.

During our visit N'kisi also kept saying, "Show the psi." "Psi" is the word he uses to refer to telepathy or psychic phenomena. And so I watched a video of an experiment designed by Rupert Sheldrake and Aimee in which Aimee is in a downstairs room, with the door closed, where a video camera (number 1) shows her actions. N'kisi is alone upstairs in his cage, with a second video camera (number 2), synchronized with Aimee's, trained on him. The two images of Aimee and N'kisi appear simultaneously in a split image on our screen. Camera 1 shows Aimee opening a sealed envelope in which an independent party has placed a picture of some flowers—which she now sees for the first time. Aimee looks at the picture; almost simultaneously N'kisi starts talking: "You gotta go get the camera, put some flowers on now. . . . You go put the pictures of flowers on there. . . . I gotta put some picture, flower. . . . Look at the little flowers, yeah . . ." A second envelope contains a photo of a man talking into a cell phone. "Whatcha doin' on the phone," says N'kisi, twice.

It is clear that a new and exciting field of research is opening before us. N'kisi's accomplishments are amazing, but will be received with scornful skepticism by mainstream science—as are many of Rupert's experiments with dogs and their owners. Aimee is seeking funding to install a surveillance system that will accurately recall everything N'kisi says and much of what he does. Then, perhaps, at least some of the skeptics will be humbled. When I think of N'kisi's astonishing abilities, I cannot help but think about the thousands and thousands of parrots around the world, captive and alone, imprisoned in tiny cages that are often too small for them to stretch their wings. We need to open our eyes and value each animal for what and who he or she is.

How amazing are the inhabitants of these animal worlds around us! Even when complex behavior is dictated by instinct alone—when it is innate or hard-wired through genetic inheritance—it is wondrous. But when we also admit that, in addition, many species of animals, certainly mammals and birds, have personalities, minds, and feelings, we are humbled indeed. And when we accept that even those with less well developed brains probably experience pain, or at least unpleasant sensations, we shall begin to think about our relationship with other animals in a new light. Not only will we be humbled—we shall also be shamed. Because our thoughtless ignorance has resulted in so much cruelty.

WHAT IS IT LIKE TO BE ANOTHER ANIMAL?

In formulating this Trust we are suggesting that you try to imagine living as an ape, a dog, a dolphin, a bat, a pigeon—even a worm. Or an earwig. And I mention an earwig simply to stretch your imagination and perhaps bring a smile to your face; a German scientist has just discovered that it possesses two *functional* penises. The sensory world of so many animals is not only rich, but so very different from ours. Some, like dogs and pigs, rely heavily on their sense of smell. Bats and dolphins use echolocation. Others, like eagles, vultures, and baboons, have incredible eyesight. Their sensory abilities can be immeasurably superior to ours. Can you hear a dog trainer's whistle?

The less like us an animal is, the harder it becomes to imagine living in its world. But we can assume that earthworms would prefer not to be trodden on and, like Albert Schweitzer and St. Francis of Assisi, we can rescue them off the sidewalk. Any of us who have shared our lives with dogs, cats, rabbits, horses, canaries, or other animals develop a good feeling for how they see or sense the world around them. If, that is, we treat them as companions, individuals possessing their own personalities, likes and dislikes, and do not regard them as possessions, as things that we own and dominate.

This Trust urges us to open our eyes and our minds to treat animals in ways we believe will be best for them in their own worlds. If we all lived by the spirit of this Trust it would be a wonderful world. Literally billions of human and animal beings would be spared untold misery and suffering. We would see animals as teachers, not chattel or inanimate objects, and we would treat them accordingly. We would not allow the brutal and unspeakable actions that are perpetrated against animals around the world.

It is so important to realize that the life of each individual animal matters in the scheme of things. Most people, even if they do believe, deep down, that this is true, put the thought out of their minds. They live in constant denial. The vast machine of propaganda, the advertising and window dressing, ensures that the general public is not faced with the consequences of their meat-eating habits, their shopping choices, their way of life. A typical response, if you try to tell someone what really goes on behind the scenes on a factory farm, in an abattoir, in a medical research lab, or in some other place of animal suffering, is, "I really don't want to hear about it. I hate cruelty" or "I love animals. I am too sensitive to listen to this." And, too, they are afraid of being made to feel guilty.

Another response I often hear is, "But those animals are *bred* for food/research/whatever." In other words, they are not like "real" animals. In some ways this is true, for animals bred into the misery of tiny solitary cages or into conditions of horrendous overcrowding never have the opportunity to live like normal individuals of their species. But unfortunately these animal slaves are still born with their ability to suffer pain and frustration and fear intact. Just like the human slaves who were born on the plantations.

Reading about some of the circumstances to which animals are actually exposed makes us feel deep shame for our own species. Here are two descriptions of research projects. The first involves dogs used in studies of learned helplessness, studies that have not been very helpful for learning about human depression, one of the reasons for which they were conducted. The second involves monkeys used in radiation research. You be the judge.

"When a normal, naive dog receives escape/avoidance training in a shuttle-box, the following behavior typically occurs: At the onset of electric shock the

dog runs frantically about, defecating, urinating, and howling until it scrambles over the barrier and so escapes from the shock. . . . However, in dramatic contrast . . . a dog who had received inescapable shock while strapped in a Pavlovian harness soon stops running and remains silent until the shock terminates. . . . It [sic] seems to 'give up' and passively 'accept the shock.'"

"In one set of tests, the animals had been subjected to lethal doses of radiation and then forced by electric shock to run on a treadmill until they collapsed. Before dying, the unanesthetized monkeys suffered the predictable effects of excessive radiation, including vomiting and diarrhea. After acknowledging all this, a DNA [Defense Nuclear Agency] spokesman commented: 'To the best of our knowledge, the animals experience no pain.'"

It seems to me that this spokesman's "knowledge" is merely wishful thinking, molded, perhaps, by a guilty conscience. How do you think the dogs and monkeys felt in the unforgiving, incomprehensible, and brutal worlds in which they found themselves? Try to imagine what it would be like to be them, to be nonconsenting victims of such unspeakable abuse.

A few months ago I watched a film sequence of a cougar hunt. I did not realize cougars, or mountain lions, could legally be hunted with dogs. I learned from wildlife photographer Tom Mangelsen how "outfitters" make money out of this. The dogs are fitted with collars especially designed to give signals when they start baying, when they have found their quarry, and chased him or her up a tree. Often, after setting his dogs on a trail—after a fresh snowfall, which gives optimum conditions for tracking—the outfitter goes home. He calls his client when the dogs start baying, for they are rigorously trained to hunt cougars and only cougars. The client may take a day or two to arrive. The cougar—and the hungry dogs—must wait. The client sometimes does the "sporting" thing and rides a horse to the kill, but often is driven up in a truck or snowmobile. The client shoots the cougar with a rifle—or to be more "manly" with a handgun or bow and arrow.

The film I watched was made by Cara Blessley Lowe, after she persuaded an outfitter to take her along on a hunt. I cannot get the images from my mind. Five dogs are baying at the base of the tree, lunging up at the trunk, frantic for action after their long wait. Up in the tree, as high as she can get, the cougar is crouched—petrified—a thin drool of saliva hanging from her lips. In agonizingly slow motion she shifts her weight slightly. Her eyes, her beautiful cat eyes, are wide with fear.

"When this is all finished," says the outfitter (meaning the shooting and stuffing), "it" (meaning the beauty of the cougar) "will last forever." Suddenly, shockingly, the gun goes off and the cougar crashes to the ground. Only her body remains.

When I studied coyotes in Wyoming, I lived on a ranch owned by an outfitter. People, predominantly men, would pay the outfitter upwards of two thousand dollars for an opportunity to shoot a black bear. Many of these people were grossly out of shape and could not follow the outfitter into the wilds. They were also clumsy, and bears could escape when they heard the hunters coming. Often they'd realize their goal by shooting a "baited" bear who came to feed at a trash dumpster, despite the outfitter's disdain for this behavior. But he'd been paid a fortune by these men who wanted to shoot a bear and weren't going to return home without a trophy and a tall tale to tell their friends.

How is it that we humans can be so brutal, so unkind, so seemingly unfeeling to so many animals despite our large brains and our ability to empathize and love? Is it because we are truly cruel? Or is it because we do not understand the worlds of the animals we torment?

One thing must be clearly stated: animals also can be, and often are, brutal and savage to each other. The death inflicted by predators on their prey can be long drawn out and horrifying to watch. When I was observing wild dogs and hyenas on the Serengeti, I had to console myself with the thought that the victim was almost certainly in a state of shock and felt little or nothing. But, however gruesome acts of predation may be, those actions are dictated by the need to survive. And this is what has characterized the relationship between humans and animals through much of our time on Earth. How different the image of a prehistoric hunter, armed only with bow and arrows and a spear, stalking his prey despite the presence of powerful carnivores. Today modern technology enables us to control, dominate, torture, and kill even the most powerful animals, as well as inflict unspeakable torture and wage war on each other.

Animals, too, show behavior resembling primitive warfare. Chimpanzees may attack members of neighboring communities so savagely that the victims die of their wounds. Chimpanzees have even developed a concept of "in group, out group" and subject those "others" to the kind of aggression almost never seen in fights between community members—twisting limbs as though trying to break them off, tearing at the skin, drinking blood. They treat the "other" like a prey animal. Hyenas have brutal territorial fights very similar to those of chimpanzees.

The knowledge that other animals can be aggressive and that we humans probably shared a common ancestor with chimpanzees five or six million years ago, although suggesting that we have inherited aggressive tendencies from our ancient heritage, does not excuse our cruel behavior today. We have developed a more sophisticated brain and are thus better able to understand the suffering of others. Indeed, despite the bad picture of humanity that is painted daily by the sensation-loving media, most people, at least most of the time, like most other animals most of the time, show more caring and nurturing behaviors than aggressive ones and mostly get along with each other reasonably well.

ANIMALS CARING FOR ONE ANOTHER

In chimpanzee society, family relationships are close and enduring. Not only do mothers rush to the defense of their infant and juvenile offspring, but they also try to assist fully adult sons and daughters. Offspring often help their mothers, and siblings frequently support one another. An orphaned infant will be cared for not only by an older sister, but by a brother too. Provided that the child is able to survive without the mother's milk, his or her life may be saved. A most touching example is the story of Mel, who lost his mother when he was three years old. He had no brother or sister but was, to our amazement, adopted by a twelve-year-old male, Spindle, who was certainly not closely related and had rarely spent time with Mel and his mother. Yet Spindle carried Mel from place to place, shared his food, and drew him into his nest each night. He even rushed to gather up his little charge when Mel got too close to socially roused males and was in danger of being dragged or hurled as a prop—instead of a branch or a rock—during charging displays. Spindle definitely saved Mel's life.

There are many accounts of chimpanzees in captivity risking their lives to try to save companions who have fallen into the water of the moat around their exhibits—for chimpanzees cannot swim. During an aggressive interaction between three adult males, one tried to escape by running into the shallow water, where he became entangled in thick weeds. The more he struggled, the more hopelessly he was caught. Amazingly, his erstwhile aggressors then ran to try to rescue him, pulling at his arms in a vain attempt to save him. Another time, in another zoo, an adult male actually fell in the water and lost his life as he attempted to rescue a drowning infant.

There are many heartwarming and heartbreaking stories that, taken together, give us real insight into the way animals may care for each other—and care for us humans, too. These stories are important because they provide yet more evidence for those of us seeking, always, to underscore the importance of individual animal lives.

Two beautiful malamutes, Tika and her long-time mate, Kobuk, had raised eight litters of puppies together and were now enjoying their retirement years. Kobuk was charming and energetic and always demanded attention. He always let you know when he wanted his belly rubbed or his ears scratched. He also was quite vocal and howled his way into everyone's heart. Tika was quieter, pretty low-key. If anyone tried to rub Tika's ears or belly, Kobuk shoved his way in. Tika knew not to eat her food unless it was far away from Kobuk. If Tika happened to get in Kobuk's way when he headed to the door, she usually got knocked over as he charged past her. Tika and Kobuk learned to work out their differences over the years they were together.

But things between Kobuk and Tika were about to change. One day a small lump appeared on Tika's leg, diagnosed as a malignant tumor. Overnight Kobuk's behavior changed. He became subdued and would leave not Tika's side. After she had to have her leg amputated and was having trouble getting around, Kobuk got worried about her. He stopped shoving her aside and did not even mind if she was allowed to get on the bed without him.

About two weeks after Tika's surgery, Kobuk woke up his human companion, Anne, in the middle of the night the way he did when he really needed to go outside. Tika was in another room, and Kobuk ran over to her. Anne got Tika up too and took them both outside, but they just lay down on the grass, Anne heard Tika whining softly. She saw that Tika's belly was huge and swollen. Tika was going into shock, so Anne rushed her to the emergency animal clinic in Boulder. The veterinarian operated on her and was able to save her life.

If Kobuk had not fetched Anne, Tika almost certainly would have died. Tika recovered and as she got healthy, Kobuk became the bossy dog he had always been, even as Tika walked around on three legs. But now Anne knew that Kobuk would be there for Tika if she needed him. Tika had known this all along. They were love dogs doing for each other what needed to be done.

Some scientists have tried to explain acts of courage and self-sacrifice in animals through two theories. In the first, kin selection, the one who is helped is related to the helper. In this case, the theory goes, the two share some proportion of their genes and the helper, it is suggested, helps the other in order to ensure the survival of his or her genes. The second theory, reciprocal altruism, is used to explain why an animal helps an individual who is not related—the helper acts in the "expectation" that he or she will, in turn, be helped. "You scratch my back and I'll scratch yours." And so, it is argued, even the noblest-seeming behavior is biologically dictated by the selfish genes of the helper. But although the two theories, kin selection and reciprocal altruism, have undoubtedly played a part in the evolution of altruistic behavior, many animal species, in addition to human beings, seem to have moved way beyond. Think back to the guide dog in the World Trade Center who, having been set free, returned to help, indeed save, his human companion. Do either of these theories explain such an act of compassion and caring? We need to see our animals in a larger light.

JETHRO, A BUNNY, AND A BIRD

Jethro, my companion dog for more than a decade, is low-key, gentle, and well mannered. He has never chased animals who live around my mountain home, and he just loves to hang out and watch his animal friends. Twice he went out of his way to be nice to two small animals who needed care. Whether Jethro expected something in return for his acts of kindness cannot be known, but it seems unlikely.

One day, when Jethro was about two years old, I heard his footsteps on the porch. Instead of whining as he usually did when he wanted to come in, he just sat there. Through the glass door I noticed a small furry object in his mouth.

My first reaction was, "Oh no, he killed a bird." But when I opened the door, Jethro dropped at my feet a very young bunny—drenched in his saliva—who was still moving. I couldn't see any injuries. The bunny was just a small bundle of fur who needed warmth, food, and love. I named her Bunny. I guessed that Bunny's mother had disappeared, probably eaten by a coyote, red fox, or mountain lion. Jethro looked up at me, wide-eyed, looking for praise for being such a good friend to the bunny. He was so proud of himself. I patted him on his head, rubbed his tummy, and said, "Good boy." He liked that.

When I picked Bunny up, Jethro got very agitated. He tried to snatch her from my hands; he whined and followed me around as I gathered a box and a blanket. I gently placed Bunny in the box. After a while I put some water, mashed up carrots, celery, and lettuce near her, and she tried to eat. All the while, Jethro just stood behind me, panting, dripping saliva on my shoulder, and watching my every move. I thought he would try to snatch Bunny or the food, but he just stood there, fascinated by this little ball of fur slowly trying to get oriented in her new home.

When I had to leave the box, I called Jethro to come with me, but he simply wouldn't come. He usually came immediately, especially when I offered him a bone, but now he stayed near the box for hours on end. When I tried to get Jethro to go to his usual sleeping spot, he refused. I trusted Jethro not to harm Bunny, and he did not during the two weeks I nursed her back to health. Jethro had adopted Bunny. He was her friend. He made sure that no one harmed Bunny.

Finally, the day came when I introduced Bunny to the outdoors. Jethro and I walked to the side of my house, where I released Bunny from her box and watched her slowly make her way into a woodpile. Bunny was very cautious. Her senses exploded with new sights, sounds, and odors. Bunny remained in the woodpile for about an hour until she boldly stepped out to begin life as a full-fledged rabbit. Jethro remained where he had lain down and watched the whole scenario. He never took his eyes off Bunny and never tried to snatch her.

Bunny was Jethro's friend, and he hoped to see her once again. Whenever rabbits come near my house, Jethro looks at each of them, perhaps wondering if they are Bunny. He tries to get as close as he can, but never chases them.

Jethro truly loves other animals. Nine years after he met Bunny and treated her so kindly, Jethro came running up to me with another wet animal in his mouth. Hmm, I wondered, another bunny? I asked him to give it to me, and he did. This time the wet ball was a young bird who had flown into a window. It was stunned and just needed to regain its senses. I held it in my hands for a few minutes. Jethro, in true fashion, watched our every move. When I thought it was ready to fly, I placed the bird on the railing of my porch. Jethro approached it, sniffed it, stepped back, and watched it fly away.

Jethro saved two animals from death. He could easily have gulped each down with little effort. But you do not do that to friends, even if they will never be able to pay you back in the future.

HELPING OTHER SPECIES

Many other true stories tell of animals helping individuals of other species. Once, in the Serengeti, Hugo and I saw pelicans form a circle about a heron who had been attacked and forced into the center of a small lake. A large bird of prey repeatedly swooped down to try to reclaim his victim from the water, but the pelicans, spreading their huge wings, drove him off and eventually the attacker gave up. We rowed out and rescued the heron from the water, and after a while it recovered and flew away.

Predators and prey sometimes become friends. In Samburu National Park in Kenya, an adult lioness was reported to have adopted a baby oryx, not only providing affection and attention but actually allowing the mother oryx to feed her calf. The baby oryx was killed by a male lion while the lioness napped, and staff at the Samburu Lodge reported that she became angry at the male lion.

Vincent Kapeen, a nature expert, noted that the lioness may have adopted the baby oryx because she had lost the company of her pride and was lonely. A few weeks later the same lioness adopted another young oryx and again allowed the newborn to feed from his mother. And a few weeks later a third oryx was yet again adopted by the lioness.

Then there was Mooney, a purebred Andalusian horse, who was trapped up to her neck in thick mud after falling into a river. There she would almost certainly have died, if her owner had not been alerted by the loud, insistent braying of Tricia, the donkey, who had been her friend for twenty-three years. Eventually, after a hard struggle, fire fighters managed to rescue the horse.

ANIMALS AIDING HUMANS

Thousands of stories tell of dogs rescuing members of their human families—or even complete strangers. A stray dog was credited with saving the life of a freezing baby in Romania. In Portugal a five-year-old boy was kept alive by dogs who lay on top of him during a rainy and windy night in the mountains. In Chile, a ten-year-old boy lived in a cave with stray dogs after he was abandoned by his parents.

Here are a few amazing stories of dog heroes who saved their human companions from serious injury or death. One freezing winter day, sixty-one-year-old Jim Gilchrist was out walking his two dogs, Tara, a rottweiler, and Tiree, a golden retriever. He decided to head home across a frozen lake. Suddenly the ice cracked, and he fell through. Tara, hearing his calls, rushed over—but fell through the ice herself. As they struggled in the icy water, Tiree appeared. Whining loudly, she crouched on her belly and crawled slowly toward them. Gilchrist reached for her collar, and this enabled Tara to scramble onto his back and out of the water. There she also lay on her belly alongside Tiree so that Gilchrist could grab her collar with his other hand. Then, as he hung on, the two dogs clawed their way backward until all two hundred pounds of him was out onto the ice. "They risked their lives to save me," he said.

It was on a balmy summer day that the two-year-old Sean Harry, who was playing among the pecan trees in his grandmother's backyard, suddenly began screaming in terror. Haven, his grandmother's Chihuahua, raced over to him and grabbed onto the three-foot-long black water moccasin that was hanging onto the seat of his jeans. Haven put her paws on Sean's leg and shook the snake until she got it off. Fortunately, it had had enough and slithered away. Sean's jeans were sprayed with venom, but he had not been poisoned—a bite would have killed him and would have killed Haven as well.

On the other side of the world, in Tanzania, a friend of mine, Dimitri Mantheatis, was also saved from a snake by his dog. It happened when he went into the garage to get his car. He was annoyed to hear a hissing sound indicating, he thought, a punctured tire. As he bent down to investigate, his dog Terry growled and jumped up at him. Pushing the dog away, he once more stooped to peer under the car. This time Terry nearly knocked him over, barking loudly. Realizing something was wrong, Dimitri stood back and switched on his flashlight. Under the car was coiled a very large, lethal puff adder.

Norman is a blind yellow lab. He loved to walk with his human family along the water, usually staying close by. But one day he heard the screams of two children and ran off in their direction. Fifteen-year-old Liza and her twelve-year-old brother, Joey, had got caught in the river's current. Joey had managed to reach the shore, but his sister was struggling, making no headway, and in great distress. Norman jumped straight in and swam after Lisa. When he reached her, she grabbed his tail, and together they headed for safety. Norman's photo hangs in pride of place on Lisa's bedroom wall. She thinks of him as her guardian angel. Indeed, a number of people rescued from trouble—such as the little girl lost at night in the snow, the man who'd fallen in a remote place and broken his legs, the young woman who suddenly found herself confronted by three youths on a dark lonely street—firmly believe that the dogs who appeared to help them, out of nowhere, were their guardian angels.

Other animals also have been reported to save humans. Dolphins of different species rescue humans in difficulties time and time again. Often these rescues involve driving off sharks who are attacking a human victim. Injured humans are kept at the surface of the water and gently nudged toward shore. There is even a story of a female bear caring for a sixteen-month-old toddler who was found safe in the bear's den in the Loreston province of Iran.

LuLu was a pot-bellied pig who lived with Jo Ann in her bungalow close to a road. One day LuLu did something she had never done before—she broke through the fence at the end of the backyard and stopped in the middle of the road. A passing motorist observed several cars swerve around her, but he stopped and got out to see what was wrong. At once, LuLu got up and walked back through the fence, then stopped and looked at him. Intrigued, he followed and found that Jo Ann had collapsed, having suffered a heart attack. Quickly he called an ambulance: the doctor said that if help had come any later, Jo Ann would almost certainly have suffered permanent damage.

Cats also can be heroes. Bernita was reading in the sitting room, and her six-week-old baby, Stacey, was asleep in her cot in the next room. Their cat, Midnight, began to behave in a strange fashion, repeatedly jumping onto Bernita's lap and hitting at her legs. Bernita pushed him away, but he persisted. Then he left the room. Suddenly Bernita heard him moaning and growling through the baby monitor. Agitated, she rushed into the baby's room, where the frantic cat was perched on a cupboard calling out and staring at Stacey—who was blue and struggling to breathe. But for Midnight, Stacey would have been just one more crib death.

Another true story tells of a pet rat, Whiskers, who slept in the kitchen with the Smiths' family dog. One night, the house began to fill with smoke. The dog remained quietly in the kitchen, but Whiskers climbed up the stairs. Mrs. Smith heard him scratching and scraping at her daughter's door. Then he began scratching outside her room and squeaking. Unable to ignore the sounds any longer she went to the door—and smelled the smoke. Quickly she woke her husband and her daughter. They were able to get out of the house and call the fire brigade—just in time. Of course they took their animals with them.

Daisy, a dairy cow, has a close relationship with her farmer, Donald Mottram, and comes whenever he calls, bringing the herd with her. One day, the farmer was savagely attacked by a newly arrived bull. He was down on the ground; the bull was goring him and stamping on his back and shoulders. At this point, severely injured, he fainted. When he regained consciousness, Daisy, who must have heard his screams, had arrived with the rest of the herd. They had formed a protective circle around him. The furious bull was on the outskirts, and though he repeatedly tried to break through, the cows drove him off. And they maintained their ring around him as he dragged himself home. When asked why he thought the cows had protected him, he said, "I have treated the animals reasonably, and they have looked after me in return. People say I am too soft, but I believe you reap what you sow."

Our primate relatives have also been known to save human lives. An eight-year-old female western lowland gorilla made headline news when she rescued a three-year-old boy who fell eighteen feet into her enclosure at the Brookfield Zoo in Chicago. Binti Jua, Kiswahili for "Daughter of the Sun," approached the unconscious child and cautiously touched him. Then she gently picked him up off the concrete and, with her own infant on her back, cradled the human child in one arm and carried him to the door to her bedroom quarters. There she sat, quite calmly, until she heard a keeper on the other side of the door. She walked away, allowing the child (who soon recovered) to be retrieved. The water hoses that had been readied—for the other gorillas were showing signs of agitation—were not needed. A full-grown male gorilla at the Jersey Zoo also rescued a child who had fallen into his enclosure, keeping the other gorillas away.

Old Man (who was given this name as an infant because depressed chimp infants do indeed look old) was shipped to North America from Africa when he was about two years old, having been torn from the body of his dying mother. For the next ten years of his life he lived in a typical five-by-five-foot cage and was used in cancer research. He was fortunate—when he was about twelve years old he was retired onto an island surrounded by a moat at a zoo in Florida. He lived with three females, one of whom presently gave birth. When Marc Cusano was employed to look after them, he was told to keep well away: "They hate people and will try to kill you."

Over the next weeks, as he watched them, he saw how humanlike their behavior was, how they flung their arms around each other with cries of joy as he approached with their food, and how Old Man was so protective of his little son. "How can I care for these wonderful creatures," Marc thought, "unless I have some sort of relationship with them?" And so he went closer and closer in his little boat. One day Old Man took a banana from his hand. One day Marc dared to step onto the island. One day, he dared to groom Old Man—and Old Man groomed him in response. And one day, they played. It was some weeks later that Marc slipped on muddy soil and startled the baby, who screamed. Instantly the mother raced over and bit Marc's neck. The other two females ran to help—one bit his leg, the other his wrist. Suddenly Marc saw Old Man thundering toward them, hair bristling in rage, coming to protect his precious baby, Marc thought. Old Man, however, pulled the females off his human friend and kept them away, while Marc dragged himself to his little boat and safety.

For me, this is a very symbolic story. Old Man was a chimpanzee who had been abused, horribly abused, by humans, yet he reached out to help a human friend in a time of great need. Surely we, with our more sophisticated brains, our greater capacity to understand suffering, can reach out now to those animals who need us so desperately?

HOW ANIMALS AND NATURE CAN HELP US HEAL

Finally, we come to the healing power of our relationships with animals. Dogs, in particular, provide tremendous therapeutic benefit to the old, the sick, and the disturbed. Other kinds of animals can be hugely beneficial too, especially cats and dolphins. Mentally disturbed patients become calm if they sit watching the fish in an aquarium. And thousands of us know the peace that can steal through us if we can spend a little time among trees and flowers, in a meadow or garden or forest, especially when birds are singing. We have a deep psychological need to connect with nature and with animals. As Anatole France said, "Until one has loved an animal, a part of one's soul remains unawakened."

I have had letters from people in different parts of the globe who survived horribly abusive childhoods only because they were able to form an attachment to some animal. One child, who was sexually abused night after night, crept down to the cellar, afraid and hurting, and cried herself to sleep with her arms around their dog. Her father found out and killed the dog. She wanted to kill herself, she told me, but on her way home from school she passed a field of cows, and one of them eventually let her come close. She went every day until they were so used to her that she could press close and relax as they lay chewing their cud. With them her tortured spirit found some healing.

And there is the extraordinary story of Misha, who, after escaping from a house in Belgium at the age of thirteen (her parents had already been taken by the Nazis) at the start of World War II, walked some thirty thousand miles through the forests of occupied Europe. Several times she formed bonds with homeless dogs and was twice accepted by wolves—once for a short period by a pair (until the male was shot), and once for several months by a family who even allowed her to share their food.

At a correctional facility in Goochland County, Virginia, inmates are allowed to train dogs in partnership with a local organization called Save Our Shelters. Inmates report that the dogs become part of their family, sharing food and toys. They also stress that the dogs bring humanity to the otherwise bleak atmosphere of the facility. Homeless people also bond closely with dogs. At the Union Gospel Ministry in Seattle, Washington, Dr. Stanley Coe organized the Doney Memorial Pet Clinic to care for the pets of the homeless people. For many of the homeless people, their companion animals are the only friends they have. A Mr. Wright says it so well, "She is my best friend and companion. . . . She will not turn her back on me like all the others."

In his speech to the U.S. Senate about slaughterhouses (Second Trust), Senator Byrd also spoke about the importance of companion animals—our "unselfish friends"—for our own well-being. It is well known that dogs can help to reduce stress in children and adults. Dreamworkers, an Atlanta-based therapy-animal group, cannot keep up with the demand for its animals by humans who need them. In fact, there is a reciprocal relationship. Touching and petting a dog can be calming, for both the human and the dog. Marty Becker has written a wonderful book titled The Healing Power of Pets, *which shows how pets can keep people healthy and happy—they can help to heal lonely people in nursing homes, hospitals, and schools.*

In his book Kindred Spirits: How the Remarkable Bond Between Humans and Animals Can Change the Way We Live, *the holistic veterinarian Allen Schoen lists fourteen ways in which a relationship between animal companions and humans can reduce stress. Some of these are reduction in blood pressure, increase in self-esteem in children and adolescents, increase in the survival of victims of heart attacks, improvement in the life of senior citizens, aiding in the development of humane attitudes in children, providing a sense of emotional stability for foster children, reduction in the demand for a physician's services for nonserious problems among Medicare enrollees, and reduction in the feeling of loneliness in preadolescents. Bringing pets to the workplace can reduce stress, improve job satisfaction, foster social interactions, and increase productivity.*

Dogs as well as cats, llamas, and dolphins also help in treating humans suffering from terminal illnesses, interminable pain, and severe dementia by providing "creature comforts." Caregivers and patients report that animals are "safe" and provide relaxation, friendship, and "bundles of love" to people in need. Many heartwarming stories are reported by Michelle Rivera in her book Hospice Hounds, *which begins with Michelle's personal story of her own mother's illness and request for the company of a dog. On her deathbed, Michelle's mother, Katherine, pleaded: "I need to see a dog. I need a wagging tail." But circumstances were such that there was no family dog at the time. Katherine had been living in a senior citizen complex that did not allow dogs. She then moved in with Michelle when she became terminal, but Michelle and her family did not have a dog because they did not have the time to properly care for a companion animal during their busy days and nights. Eventually Michelle quit her job to stay home and take care of Katherine, so the time was right to bring a dog into their lives. Tyrone joined the family and was there when Katherine died, along with Sable, Michelle's cat. Katherine's wish had been fulfilled and her two loving buddies, healers with fur coats, had kept company on the bed beside her, providing much-needed companionship.*

My mother, Beatrice, suffers from dementia. Once, while I was visiting my parents, my father called his friend, Ginger, whose husband had recently died, so that she could show me her new treasure, a teacup poodle not surprisingly named Tiny, whom she carried inside her shirt! Tiny, whom Ginger pampered and deeply loved and who also pampered and deeply loved Ginger, brought Ginger much joy in the absence of her husband. Unfortunately Ginger had to move from her home because the rules imposed by the homeowners' association forbade dogs. I can guarantee you that this wonderful small dog was much less of a nuisance than most of Ginger's human neighbors, yet Ginger had to move because dogs were banned. What was very interesting to me was that my mother, who had been bitten by a dog when she was young and feared dogs throughout her life, also found Tiny to be a welcome and comforting friend. She actually let Tiny lay in her lap and smiled from ear to ear as Tiny burrowed into her blanket and heart.

Why is this so—why are dogs such good healers? For one, if we allow dogs into our lives, they readily awaken our senses, spirits, and souls. They, and many other animal beings, offer us pure and unfiltered respect, humility, compassion, trust, and love. We have much to learn from them.

THE FOURTH TRUST

TEACH OUR CHILDREN TO
RESPECT AND LOVE NATURE

The Fourth Trust concerns the need to provide the right experiences for our children as they grow up. Their attitudes toward animals are shaped as they learn from those around them, those they love and admire. It is desperately important that we nurture new generations of young people who will care for nature and help to heal some of the hurts we have inflicted.

Children brought up in a caring family that respects animals, especially when those children grow up in close contact with some animal, tend to be kind to animals as adults and to be loving and compassionate individuals. When children from seemingly normal homes show extraordinary cruelty toward animals, it may indicate some serious psychological disturbance. When the childhood of mass murderers, serial killers, and other psychopaths could be traced, it was often found that there was a history of horrible cruelty toward animals. Children's relationships to animals, then, can be a means of teaching compassion and kindness or, alternatively, an indicator, when those qualities are lacking and abuse of animals is present, of potential psychological problems.

I am fortunate in that I had an extremely wise and supportive mother who nurtured my early passion for all life and allowed me to surround myself with a variety of animals. Once when I was sixteen months old, she came to say goodnight and found me watching, absolutely entranced, something on my pillow. Coming closer, she saw a small collection of wriggling earthworms. Instead of scolding, she told me quietly that they'd die without earth. I gathered them up quickly, she said, and toddled with them to the garden. Even when we lived in a small apartment in London we shared our space with a dog—a wonderful bull terrier called Peggy. Later we moved to a house near the sea where we had our own garden, or backyard. There, my sister and I cared for a series of dogs and cats, guinea pigs who went for walks on harnesses, a canary who used his cage only for feeding and sleeping, a hamster who made her nest in the upholstery of an armchair, and all manner of other creatures, none of whom were caged.

I learned from nature, spending all my free time outside. When I was four years old, we went to a farm for a holiday. There I helped to collect the hens' eggs. There were no cruel battery cages in those days. I could not work out where the egg came from—where was there a hole that size? So I hid in one of the henhouses to find out. For four hours. Again, despite her concern (the family had called the police), my mother, when finally she saw me rushing to the house all excited, sat down to hear the wonderful story of how a hen lays an egg. I also learned from the books that my mother found for me about animals. I read and read about animals. Doctor Doolittle and Tarzan and Mowgli. One of my favorite books (which came free after my mother had saved countless coupons from something like soapflake packets) was *The Miracle of Life*. As I mentioned before, this book, with extensive text illustrated by countless black-and-white photos and line draw-ings, dealt with topics as diverse as camouflage, "many tongues for many purposes," insect pests, and the history of medicine. It was writ-ten for adults, but I loved it. It encouraged my curiosity and my desire to know more and ever more about the natural world.

I developed my ideas about "minding animals" early in life, according to my parents. "Minding animals" means two things. First, "minding animals" refers to caring for other animal beings, respecting them for who they are, appreciating their own worldview, and wondering what and how they are feeling and why. The second meaning relates to the fact that many animals have very active and thoughtful minds.

I did not grow up in the country; nor was I raised with animals. I lived with some goldfish when I was a youngster and always wondered how they liked being in their fish tank. I always felt what Rachel Carson called "a sense of wonder," an insatiable curiosity about the world in which I was immersed. My home was filled with compassion and love. My parents told me that I always wanted to know what animals were thinking and feeling. They recalled that when I was about four years old, I yelled at a man for hitting his dogs, and the man chased after my father! My father remembered that when we were on a cross-country ski trip in Pennsylvania when I was five years old, a red fox ran in front of us as we broke trail and I was absolutely taken with the beauty of the animal. I immediately asked questions about where the fox lived and whether he was happy. All of my interactions with animals were helpful in bonding me to them, and all involved loving rather than harming them. That compassion begets compassion is obvious. Years later, I dropped out of two graduate programs because I did not want to harm or "sacrifice" (kill) animals as part of my education.

Both Marc and I were lucky. There are millions of children who do not have understanding parents. And millions more are born into abject poverty, with little or no opportunity to explore the natural world. And, too, there are those born into cultures that have little regard for animal life—or for human life either. How can we teach such children about the beauty and wonder of nature? In the developed world there are organizations that take groups of inner-city kids out of the concrete jungle and into nature. But only a tiny fraction of youth is affected. There are conservation organizations in the developing world that take busloads of children into national parks or other wilderness areas. But children living in impoverished rural areas have almost never had an opportunity of this sort. What can we do about this?

One way of introducing animals to disadvantaged children is to take the animals into the schools. There was a time when I was very much against this idea. because it seemed to exploit the animals. But I have come to think differently after seeing, with my own eyes, the incredible difference that meeting an animal can make in the life of a child. Of course, such a program must be really well run and the animals protected from any kind of harm. And some animals simply are not suitable. But domestic animals as well as quiet, wild animals who cannot be released into the wild and who are used to people—can serve as wonderful ambassadors from the animal kingdom. It seems worth it, indeed, when you see a child's eyes filled with wonder.

Thousands of children, all over the world, are born into dysfunctional families, imprisoned in a world of drugs, alcoholism, and violence, a world in which there is often terrible cruelty to animals. And tragically, although children can so easily learn to care for and love animals, they can learn also, from those around them, to fear, hate, or despise animals. Such children will kick a dog or torture a cat and laugh at the animal's suffering. After all, they have often been treated in the same way themselves.

How inspiring to find that in such a hopeless world it is sometimes animals themselves who can provide healing and teach love. Peter (not his real name) was a twelve-year-old boy from an inner-city area of New York State. He was convicted of many violent crimes, including, most recently, assault with intent to kill. In a last attempt to rehabilitate him, he was sent to Green Chimneys farm, a residential haven founded fifty-five years ago by Sam Ross for boys who are violent, many of whom have been in the juvenile courts time and time again. Like all new arrivals, Peter was asked to select, from a list of farm residents ranging from pigs and horses to guinea pigs, an animal to be his special companion, whom he would look after throughout his stay. Typically, these boys are initially hostile and verbally abusive, although they are not allowed to physically harm their animals. After a week or so the child usually realizes that he has found a very special kind of friend—one who will never tell on him, never let him down. And then rehabilitation starts. Soon the child will be ready to interact with another boy and then, gradually, he will become integrated into life at Green Chimneys.

Peter did not want to choose an animal. "It doesn't matter. I don't like animals," he said.

The staff consulted with each other. "Well, we think you will really like a very special rabbit."

Peter gazed at them blankly. "I don't know what a rabbit is."

This opens a whole new window into the life experience of a child like this. He was twelve years old, and he didn't know about rabbits. No one had read him the story of Peter Rabbit. No *B* for Bunny in his kindergarten alphabet book. No images of green fields and rabbits feeding and playing in the early sunshine. No trips to the countryside. And so, "I don't know what a rabbit is," he said.

They brought out a very large white rabbit. She had been through the process many times. She was experienced and gentle. She sat quietly on the bench while they walked the boy over to her. His mentor stroked the rabbit for a while, then suggested he do the same. Slowly, he reached out and touched her. He was very still. The adults present sensed that something special was happening and stepped back. For several minutes he stood there, with his back to them, stroking the rabbit. They saw him rub his eyes. Then, slowly, he turned to face them, his face closed and expressionless.

"Well, I didn't hurt her, did I?" he said. In his experience you were repeatedly hurt by those stronger than yourself—you got your own back by, in turn, hurting those who were weaker. That encounter with a rabbit was his first experience of a different way of looking at life, his stepping-stone to a brighter future.

When I am with small children, especially my three grandchildren and my two grandnephews, I feel such sadness. We have destroyed and polluted so much of Planet Earth in the sixty plus years since I was a small child. So many animal and plant species have gone forever. It is so desperately important that the children of today learn to be better stewards than we have been.

This is why I initiated the Jane Goodall Institute's Roots & Shoots program—to help children understand more about the true nature of animals and how to conserve them.

It all began in February 1991 on the verandah of my house in Dar es Salaam, Tanzania. Sixteen Tanzanian secondary-school students, from nine schools, met with me to talk about animal behavior and the environment. We discussed the chimpanzees, with their different personalities, intellectual skills, emotions, rich family lives, and political skills. The students were fascinated. Suddenly, they realized that there was not, after all, a sharp line dividing humans from the rest of the animal kingdom. This led to a new understanding of other amazing animals. We talked of elephants, and culling. We talked of conditions suffered by chickens and goats in some markets. And the students wanted to do something about it.

We decided to establish clubs in their schools. We would call ourselves Roots & Shoots, a name that symbolizes the power of youth, if informed and empowered to take action. Roots make a firm foundation; shoots seem small and weak, but to reach the light they can break through brick walls. Imagine the brick walls to be all the problems that we humans have inflicted on the planet—environmental destruction, pollution, genetic modification of foods, greed, cruelty, crime, and war. Hundreds and thousands of roots and shoots—the youth—around the world can break through and make a better world. What started as one group of sixteen-to-eighteen-year-olds has now spread to over sixty countries with some four thousand active groups from preschool through university level.

Every group chooses at least one hands-on project, in each of three areas, to demonstrate care and concern for animals (including domestic animals), for the human community, and for the environment we all share. What they decide to do depends on local problems, the age of the children, whether they are from the inner city or a rural area, and the part of the globe they live in. The guiding philosophy is gentle: the tools for change are knowledge and understanding, hard work and persistence, love and compassion, leading to respect for all life.

Students in Roots & Shoots programs around the globe have been actively engaged in a wide variety of projects that generate respect and compassion for animals and for people. These include taking responsibility for the animals with whom they share their homes, helping injured animals, working at rehabilitation centers, rescuing dogs and cats from humane shelters, and learning about both domestic and wild animals.

Some senior citizen residences and homeless shelters allow Roots & Shoots children to take animals to visit. The Roots & Shoots program at the University of Colorado holds an annual senior ball at a residence for senior citizens in Boulder. We begin it with the residents giving the chimpanzee "pant hoot"! One student regularly walked and cleaned up after a dog who lives with his elderly neighbor, which meant he was fulfilling all three Roots & Shoots requirements—he was helping an animal, a person, and keeping the environment clean. Students also learn from one another. One girl brought a hamster to class. One of her classmates grabbed him, wanting to say "hello." She was just trying to be friendly, but the hamster shrank back in fear and became withdrawn instead of his usual outgoing self. This led to a discussion about the importance of treating all animals with respect. When, at a later date, one of the students came to class with her dog, the others reminded each other not to "get in the dog's face." Instead, they waited for him to approach them when he was ready.

THE DOG IN A LIFEBOAT

Everyone who's worked in Roots & Shoots programs has been amazed at how some youngsters develope very sophisticated attitudes about human-animal interactions by the time they are four or five years old. One thought experiment in which some elementary students have engaged in one Roots & Shoots program in Boulder is called "The Dog in a Lifeboat." In this exercise, there are three humans and one dog in a lifeboat and one of the four individuals has to be removed, thrown overboard, because the boat cannot hold all of them. Generally, when this situation is discussed, most people agree that, all other things being equal, reluctantly the dog has to be removed from the boat. It is usually argued that a human will suffer more than a dog if he or she is thrown overboard and that the dog has less to lose. Also, the dog's relatives and friends will not miss him as much as will those who survive the loss of a human.

Variations on the theme can also be introduced and result in engaging and enlightening discussions. For example, perhaps two of the humans are healthy youngsters and one is an elderly person who is blind, deaf, paralyzed, without any family or friends, and likely to die within a week. The dog is a healthy puppy. Students have admitted this is a very difficult situation, and that maybe, just maybe, the elderly human might be sacrificed because he had already lived a full life, would not be missed, and had little future. Indeed, this is very sophisticated thinking, that perhaps the elderly person had less to lose than either of the other humans or the dog. All students agreed that this line of thinking was not meant to devalue the elderly human. But, in the end, the students and most people reluctantly agree that, regardless of the humans' age or other individual characteristics, the dog has to go.

The level of discussion in these exercises is always very stimulating—consideration of quality of life, value of life, longevity, losses to surviving family and friends. But what is truly inspiring is that before alternative solutions were ever discussed, students wanted to work it out so that no one had to be thrown overboard. "Why does any individual have to be thrown overboard?" they asked. "We will not do it," they said, when they were told the thought experiment required that at least one individual be tossed; they said this was not acceptable, end of story. Now these are the kinds of people in whose hands we would feel comfortable placing our future. Some ideas about how all individuals could be saved included having the dog swim along the side of the boat and feeding him or her, having them all switch between riding in the boat and swimming, taking off shoes and throwing overboard all things that were not needed so as to reduce weight and bulk, and cutting the boat in two and making two rafts. All students thought that even if the dog had to go, he or she would have a better chance of living because more could be done by the humans to save the dog than vice versa. Very sophisticated reasoning, indeed.

"I HAVE A DREAM," "I AM THANKFUL"

Other activities in which children in Roots & Shoots programs across America and around the world have partaken involve filling in the blank and drawing a picture for the following phrases: "I have a dream that _____" and "I am thankful for _____." Ellen Bilek, a teacher in Boulder, Colorado, began this project with her kindergarten class.

Responses show clear concern and respect for animals, people, and the environment we all share. Here are some examples: "My dream is that animals get enough to eat." "My dream is that I do not want my sister to get hurt." "My dream is that my grandmother is not sad that her two brothers died." "My dream is that everyone has shelter." "My dream is to meet Nelson Mandela." "I am thankful for my fruit bowl and my house." "I am thankful for Mother

Earth." "I am thankful for forests and animals." "I am thankful for my family." "I am thankful for birdhouses and love." "I am thankful for giving." The last one was accompanied by a drawing of the kindergarten student giving a homeless person a sandwich.

Many children view their dreams and the things they are thankful for as trusts that should not be violated. It is wrong, they claim, not to share food with those who need it, to allow animals to be abused, or to waste, litter, or otherwise spoil the environment.

SOWING SEEDS OF RESPECT, COMPASSION, EMPATHY, AND LOVE

In a survey, the naturalist and author Brenda Peterson found that 80 percent of children's dreams center on animals, whereas only about 20 percent of adult dreams still include animals. She suggests that perhaps we need not only wake people up about how they relate to animals, but also bring them back to their birthright as original dreamers. There can be no doubt that children's vitality and their personal and lifelong commitments that are motivated by compassion and love will make this a better world for all.

Let us teach our children to respect nature.

THE FIFTH TRUST

BE WISE STEWARDS OF LIFE ON EARTH

In the Fifth Trust, we plead for a more compassionate and gentle stewardship of Earth and point out ways in which each of us can try to live in harmony with nature, to leave lighter footprints as we travel from cradle to grave. As we humans go about our business, we must remember how our actions will affect the other animals who also call Planet Earth their home.

With our big brains and advanced technology, we have been able to dominate all other life forms. All those raised in the Judeo-Christian belief system take this for granted, for did not God give man "dominion over the fish of the sea, and over the fowl of the air, and over the cattle, and over all the earth, and over every creeping thing that creepeth upon the earth"? However, many Hebrew scholars believe the word "dominion" is a poor translation of the original Hebrew word *v'yirdu*, which means "to rule over" in the sense of a wise king ruling over his subjects with enlightened stewardship. We have only to think of the terrible harm we are inflicting on the planet, on the environment and all living things, to realize we have disobeyed the instructions laid out in Genesis 1:26. We have not been wise stewards—rather, we have used force, when necessary, to impose our will on nature and all beings less strong than we are (including other humans).

THE EFFECTS OF HUMAN OVERPOPULATION

As a result of our Western materialistic greed and arrogance, on the one hand, and poverty and desperation, on the other, we are not tending Planet Earth and her animal inhabitants as good stewards should. Instead we have despoiled the land like thoughtless conquerors. Humans have adapted to life in almost every kind of environment on Earth, from mountain heights to steamy jungles to open plains. Adaptations that would take millions of years to acquire through the process of physical evolution have been developed in just a few years through cultural evolution and technology. When the wildness of a place or the animals living there got in the way, they were mercilessly cut down, killed, banished, enslaved, tamed. And our human numbers grew and grew.

As a consequence of this human population growth, hunting, environmental destruction, and pollution, other animal species have been increasingly crowded out and their numbers depleted, often to the point of extinction. Of course, animals went extinct even in prehistoric times through causes other than human actions. We are not responsible for the extinction of the dinosaurs, who perished, along with 70 percent of all the species then on Earth, about sixty-five million years ago, long before humans appeared. The cause of their extinction is thought to have been a dramatic climatic change due to an asteroid or volcanic activity.

But in the last hundred years, humans have been the cause of untold animal losses. Global rates of extinction are the highest they have been for the past 65 million years, and many extinctions go unrecognized. The natural rate of extinction is about one species per one million species per year. Extinctions due to humans range between one hundred and one thousand species per one million species per year. About one new species per one million species is born each year. Do the math—this is not a good situation at all, for far fewer species are born than go extinct due to human activities.

In the past five extinctions it took about ten million years to restore biodiversity—now there may be no coming back because of increased rates of extinction. Humans have clearly altered the future of biological evolution. In North America alone about 235 animal species are threatened by pollution, human encroachment on their habitat, and aggressive harvesting practices. Around Puget Sound in Washington State, satellite imaging has shown a 40 percent loss of tree cover in the past two decades, which would have removed 35 million pounds of pollutants from the air per year. It is notable that the United States spent about $100 billion on health problems related to air pollution during this period of time.

The dodo was eaten to extinction. Passenger pigeons, which once flew in vast flocks that darkened the sky for over an hour as they passed overhead, were shot until there were none left. The great bison herds that used to cover the Central Plains like a blanket have gone, although a few bison still live. All manner of less well known species have vanished forever at our hands. In tropical forests, myriads of species of insects and plants are lost during clear-cutting, never to return. Many become extinct even before they are named. And now Miss Waldron's colobus monkey is gone forever from the forests of West Africa. Within ten to twenty years the great apes of Africa and Asia could be wiped out except for a few small groups in protected areas. Indeed, we could be facing the start of the extinction of very many known mammals and birds, as well the continuing extinction of hundreds of the more obscure forms of life. We agree wholeheartedly with Professor E. O. Wilson, who believes that this destruction of biodiversity is one crime for which we shall never be forgiven by future generations.

And still our numbers grow. On October 12, 1999, the world's population was declared officially to have reached the six billion mark, twice the population of 1960. It's estimated that the world's population will increase by three billion in the next fifty years. Biologist Michael McKinney found that the human popu-

lation size is positively correlated with the threat to the numbers of birds and mammals for continental (but not island) nations, and that mammals suffer more losses than birds during initial human impacts. His data are convincing—149 nations were analyzed for mammals, 154 for birds. But fish are also in trouble due to extensive exploitation by humans. In North America, fishes are up to eight times more likely to be in peril than either mammals or birds.

NATIONAL PARKS AND REFUGES

All over the world, as people wake to the reality of the vanishing wilderness areas and the animals living there, national parks and refuges have been established by governments and the private sector. We must be eternally grateful to those who had the foresight to set aside vast tracts of land for future generations—in some places, due to human population growth and "development," there is very little of the original wilderness outside the boundaries of these parks and reserves. But even animals within so-called wildlife refuges are constantly subjected to human interference. In the United States, hunting is allowed in some national wildlife refuges. Poaching also is widespread around the world. Poachers may be poor and hungry people trying to live off the land as they have for hundreds of years, or unscrupulous individuals trying to make large profits by selling ivory, skins, body parts, and so on. Because of lack of funds, corrupt officials, or political unrest—which sometimes translates to full-fledged war—it is sometimes extremely difficult to enforce wildlife protection laws.

Another problem occurs when the population of certain species within a park becomes too great for the carrying capacity of the area set aside—either from natural reproduction or, as is the case with elephants in Africa, because groups move into the "safe" areas from outside. And then, gradually, the animals begin to destroy their environment.

Then some people believe it is necessary to "manage" the animals, which often means killing them—"culling," it is called, because it sounds less brutal. But sometimes it is very brutal indeed. The footage I have seen of the killing of whole herds of elephants—except for the small calves who can be sold to zoos—is just about the most harrowing wildlife spectacle one could ever watch. The fear, the courage of individuals who approach the guns in a vain attempt to protect cows and calves, the desperation of the young ones as they huddle, bewildered and terrified, beside the bodies of their dead mothers. Surely there must be a better way?

HUNTING

Wildlife management issues are not confined to problems within the national parks. In many parts of the world deer populations have become a problem as natural wilderness areas decrease. Then, as urban sprawl penetrates ever deeper into deer habitat, people and deer come into conflict. Deer represent a hazard to people driving on the roads at night, and they are seen as pests because they eat the things people have planted in their gardens and yards. Of course, if we look at it from the perspective of the deer, it is the cars that are the hazard, and the people who have destroyed their natural foods that are the pests. Be that as it may, the time-honored solution, certainly in the United States, but in other parts of the world too, has been to issue hunting licenses. This reduces the number of deer while increasing the revenue of the authority issuing the license.

Arguments against hunting have more punch when they come from hunters themselves—people who have turned against the so-called sport. For years, Tom Mangelsen, noted wildlife photographer and filmmaker, was "one of the great defenders" of hunting. For years he argued the merits of hunting—the need to manage wildlife and how hunting helped to accomplish that goal. The overpopulation surpluses, he argued, provided "harvestable" quotas that helped to raise money that could be used to protect other animals. "Now," says Tom, "these arguments make me gag." Indeed, it is true that the national wildlife refuges are in part maintained as a result of hunters' dollars. Ironically, for many creatures, they provide no refuge. Instead, they are killing grounds where countless numbers of birds and mammals, including trumpeter and tundra swans, fall to hunters' bullets.

Recently, when I was in Vermont I found myself in the middle of a heated argument about the merits of hunting with bow and arrow. I grew up on the tales of Robin Hood and his band of ragged followers, living off the land, as they robbed the rich to feed the poor. Robin Hood's skill with bow and arrow was legendary. Today, bow hunting is often considered more "sporting" than using a gun—yet it is, in practice, one of the more crude and cruel methods of killing animals. Few hunters can compare with Robin Hood, and those few are the only ones who can dispatch their quarry quickly and cleanly. In Vermont, according to a report by Merritt Clifton, editor of *Animal People,* some three thousand deer per year are wounded and not retrieved by bow hunters. In Texas and Illinois, bow hunters leave one deer wounded for every one they kill and retrieve. By contrast, rifle hunters leave one or two deer wounded for every eighteen or nineteen killed. In his book *The Bowhunting Alternative,* Adrian Benke reports a more than 50 percent crippling rate for victims of bow and arrow. However they are wounded, by arrow or bullet, the unretrieved deer often suffer for days before dying of internal bleeding or infection.

Deer hunting has found its way onto national television in the United States. Jeep recently ran an antihunting commercial that has generated a lot of support from a wide variety of animal protection groups. In the commercial, a Jeep is shown with two deer on the top. When the driver arrives at an area where there is a "No hunting" sign, he releases the deer, who appeared to be dead, and tells them they are safe. The deer bound away.

THE PREDATORS

One of the reasons for increasing numbers of deer is the fact that the natural predators have been systematically eliminated. Wildlife Services (formerly called Animal Damage Control), a branch of USDA, has been responsible for killing hundreds of thousands of animals—"varmints," as they call them— including, in 1999 alone, over 85,000 coyotes, 6,200 foxes, 359 mountain lions, and 173 wolves. They use leghold traps, snares, explosives, poisons, and aerial gunning, all of which cause much pain and suffering. During that period, when over 96,000 predators were inhumanely killed in the name of control and management, only about 1 percent of livestock losses were actually due to predators—99 percent were due to disease, exposure to bad weather, illness, starvation, dehydration, and deaths at birth. In addition to the many predators targeted, other animals are inevitably killed or wounded, including many domestic dogs and cats. Yet the various wildlife agencies in many states continue to persecute the predators. As a result, prey animals, such as deer, increase in number, and then the agencies make money by selling increasing numbers of permits to hunters to help "control" them.

For many years wolves were fiercely persecuted in a campaign designed to exterminate them. And in many parts of their range they vanished. They were shot from vehicles and helicopters, caught in traps, and poisoned. They were blamed for killing livestock and feared as creatures who endanger human life. Similar "crimes" were attributed to mountain lions. In fact there are no reported incidents of wolves killing humans in the United States, and the only attack was from an individual with rabies. And there are very few recorded instances of mountain lions attacking people. A total of seventeen people have been killed since 1890, and of those seven were in the last ten years—as people push farther and farther into cougar country. Attacks from domestic dogs are far more numerous.

In Wyoming, a hunter can buy a license to shoot a mountain lion for just thirty dollars. And, based on a study that Wyoming Game and Fish officials themselves admitted was not sound science, the quota of mountain lions per year was raised in 2001 from two to twenty. In Wyoming, there is a six-month hunting season and it is illegal to kill a female with cubs by her side. It is also illegal to kill cubs younger than one year. However, during the hunting season, there is a 75 percent chance that a female will have cubs, and if they are not with her they will be less than four months old. Therefore there is a good chance that any female shot will put the hunter theoretically outside the law since her small cubs will, in effect, have been killed.

Only in California are mountain lions fully protected, thanks initially to the efforts of Margaret Owings. In Texas there are no regulations at all. Mountain lions, including tiny cubs, can be killed at any time of year in any way. They can be hunted with dogs, in cars, with guns or bow and arrow; they can be trapped and poisoned. Over 70 percent of Texas is private land, which makes things difficult for those fighting for more humane hunting laws. In Colorado cougars can be hunted by dogs, and in 2001 Washington State lifted its ban on hunting cougars with dogs for the second year in a row. In Oregon scientists from the Oregon Department of Fish and Wildlife are preparing to kill as many as thirty-two cougars to "study" their feeding behavior, to determine if predation, poor nutrition, or both are responsible for a decline in Oregon's elk population.

It is to be hoped that, as public awareness increases, wolves, bears, and mountain lions, along with coyotes, bobcats, and so on, will increasingly be respected for the important roles they play in the ecosystem and be perceived as precious gifts to be preserved for generations to come—not as mere "things" to be exterminated or traded for hunters' dollars.

THE SLAUGHTER OF ANIMALS FOR BODY PARTS

One of the great tragedies, on a par with the hunting of elephants for their tusks, rhinos for their horns, and countless animals for their hides, is the relentless hunting of chirus, or Tibetan antelope. Unfortunately for them, their wool can be made into shawls known as *shahtoosh*. They are made from the very fine underfleece, and chirus have been slaughtered in the tens of thousands to satisfy the demand. Each dead chiru yields only about 150 grams of the wool; many animals die to make just one shawl. And that shawl can retail in New York, London, or Paris for thousands of dollars. So, although chirus, found only in the high mountains of Tibet and the remote Ladakh area of India, are fully protected by local and international law, they are relentlessly hunted by poachers. There are heartbreaking descriptions of the devastation—forty to fifty skinned and bloody corpses lying on the snow and twenty or so fawns desperately searching for their mothers. These beautiful antelope are almost extinct.

Tragically, the illegal trade in *shahtoosh* is also contributing to the extinction of the tiger. Tiger body parts, particularly their bones, supply the international market for some traditional Chinese, Korean, and Japanese medicines. Many of these illegal products reach China through Tibet—where the couriers often barter them for *shahtoosh*, which they can sell when they get back to India. From India *shahtoosh* is then illegally exported to the wealthy consumer countries.

THE "BUSHMEAT TRADE"

Chimpanzees are efficient hunters, although meat forms no more than 2 percent of their diet. It seems likely that the hunting and killing of wild animals for food has been a traditional way of life for very many humans since the dawn of time. Until relatively recently, subsistence hunting was not a huge threat to populations of wildlife, but as human populations grew, the number of animals shot and trapped increased. Even so, the greatest threats to wildlife around the world have been, since the early twentieth century, extermination projects (such as the attempts to eliminate wolves from vast areas of North America and Europe); hunting for "sport"; the live animal trade to supply zoos, circuses, and the demand for exotic pets (exotic birds, reptiles, and reef fish suffer upwards of 80 percent mortality during shipment alone); and medical research. And, of course, the trade in ivory, rhinoceros horn, and different parts of various animals for supposed medicinal purposes.

Today, as we enter the twenty-first century, the situation is grim, and it is the *commercial* hunting and selling of wild animals for food—the "bushmeat trade"—along with habitat destruction that threatens to exterminate many animals throughout Asia, the new tropics, and across the whole of tropical Africa. In the great Congo Basin of Central Africa, the last stronghold of the African great apes, logging companies (none of them African) and, to a lesser extent, mining companies have built roads deep into the heart of previously inaccessible forests. This opens up the forests to settlement, the cutting of trees for charcoal and firewood, the setting of snares, and, worst of all, access for hunters.

Trucks travel the roads transporting hunters deep into the forest, where they camp for a few days, shoot everything from elephants, gorillas, and chimpanzees to the larger birds and rodents, sun-dry or smoke the meat, and then truck it into the towns. This trade is not to feed starving people—it is to cater to the cultural preference of many urban people for the flesh of wild animals. In the towns a piece of chimpanzee or wild antelope will fetch more than the same-sized chunk of cow or chicken. Moreover, the indigenous pigmy peoples, who have lived in harmony with their forest world for hundreds of years, are paid to hunt for the two thousand or so logging staff and their families. None of this is sustainable, and the indigenous people will suffer terribly when the loggers move on.

Unless a concerted effort is made to control this bushmeat trade, it is likely to result in the virtual extinction of all the great apes and many other species in the remaining Central and West African forests within the next ten to fifteen years. It was only relatively recently that I realized how complex the problem is—the extent to which so many of the economies of these countries are now depending on the sale of bushmeat, the difficulty of controlling the trade and of enforcing the laws of countries that are often politically unstable where there is sometimes fighting, and where there may be corruption at many different levels. There are similar problems in Asia and Latin America.

There are no easy answers, yet we must find ways to protect these amazing forests, the animals, and also the future of the people whose homelands and way of life are so badly threatened. We need to take bold and imaginative steps, and we must act now. Some conservation and environmental groups, such as the Wildlife Conservation Society and the Jane Goodall Institute (JGI), have started programs to tackle the issue. Our Congo Basin project investigates ways of providing alternative protein to villagers and logging camps—such as game farming. Hunters will need alternative ways to earn money. Urban people need a great deal of education, so that they understand the unsustainability of their present meat-eating behavior.

The Jane Goodall Institute's (JGI) Congo Basin project is working with local hunters and with those selling bushmeat in the markets, especially the women who already understand that their future is being compromised by undisciplined hunting. We are trying to persuade logging companies to incorporate the protection of animals into their code of conduct, which currently deals only with the sustainability of the trees of the forests. We hope to persuade companies in the Congo Basin and in West Africa to prohibit illegal hunting within their concessions, to ban the transport of illegal bushmeat on their roads, and to provide alternative protein sources for workers and their families.

We are talking with governments and local people about the threat that bushmeat hunting poses to the wildlife and the health of the forest on which their future depends. We are committed to increasing awareness and understanding of this crisis and are working to do so with other concerned nongovernmental organizations on the ground. It is necessary to persuade African governments to improve and enforce their own wildlife laws. And, too, governments of the developed world and international agencies such as the World Bank must find ways to ensure that foreign assistance to, and investments in, African countries are contingent on sound conservation policies. The Jane Goodall Institute (JGI) is participating in the United Nations Development Program, which addresses this issue.

Finally, Christina Ellis of the Jane Goodall Institute (JGI) is coordinating a consumer-awareness guide to allow the people of North America and Europe to make educated purchases that will help to reduce their indirect contribution to the wildlife crisis in Central Africa. An example of a company that addresses the issue directly is the Rescued Wood Bowl Company in Fort Collins, Colorado, which uses only rescued and recycled wood. Joan Roney and her dedicated group of volunteers at Rainforest Relief want to reduce the use of timber and encourage the use of such alternative materials as recycled plastics, bamboo, and certified wood materials.

COLTAN AND CELL PHONES

Connections between resource use and the decimation of animal populations are not always obvious. Consider the mining of coltan, a source of the element tantalum, which is used as a coating for many electronic devices such as cell phones and computers. Coltan can easily be mined with a shovel. And it is mined in the Kahuzi-Biega National Park in the Democratic Republic of Congo. As a result, endangered eastern lowland gorillas have experienced a 50 percent reduction in numbers over the past few years, and elephants are virtually gone. There are other materials from which cell phones can be made. Individual cell phone users can make a difference by refusing to buy a phone for which this mineral was used. Indeed, the United Nations has encouraged people to boycott companies that use illegally mined coltan—if we do not buy their products, they will not get made.

REINTRODUCTION PROGRAMS

In many parts of the world people are trying to reintroduce animals into areas where they once existed. One of the best known and most successful of these programs is the reintroduction of wolves into Yellowstone National Park. There was a great deal of opposition when the scheme was first suggested, but the process was very carefully planned to take into account the interests of ranchers, farmers, and tourists—and, most important, the needs of the wolves. The reintroductions proceeded relatively smoothly, though a few wolves were shot by angry farmers at the outset. Now the wolves are reproducing and once again these predators are fulfilling their time-honored role in the Yellowstone ecosystem—maintaining healthy populations of their prey species. Researchers are now investigating the habitat in Valais, Switzerland, to see whether European wolves might have a chance of survival in their old hunting ground. It is very important to carry out this kind of careful research—without such facts, reintroduction projects may be doomed to failure.

The story of Lucy shows what can go wrong when we try to redecorate nature. Lucy, an adult female Canadian lynx, was living in northern British Columbia when suddenly her life changed. One cold winter morning she set off to hunt for food. Maybe she would encounter a mate, but on this day, trying to avoid the humans who have invaded her territory, she stepped on a trap. Snap. Her left front leg was caught in a leghold trap, and she fell. Panicking, she struggled to free herself, and the jaws of the trap tore deeper and deeper into her leg, causing a bad injury that would make walking difficult. She was one of the lucky lynx—her leg was not broken. Humans approached and shot her with a tranquilizing dart to calm her down. They placed a radio collar around her neck, put her in a box, and loaded her in a truck.

Thus began Lucy's long journey to southwestern Colorado, where she and other captured lynx were held for a few weeks in cages before being released into the wild. Southwestern Colorado is a marginal habitat in which her favorite food, snowshoe hares, are very rare. Five days after being released Lucy starved to death. So did three other lynx immediately after being released. These individuals had been thriving in their home in British Columbia, only to be killed in human attempts to restore an ecosystem that did not have adequate food. This fact was known before the reintroduction project began. Many other lynx also succumbed to difficult deaths after being moved from British Columbia to Colorado. As of March 2002, at least forty of ninety-six reintroduced lynx had died, including nine who starved to death, five who were shot, and five who were struck by cars. There hasn't been any breeding, the absence of which spells doom for the future of lynx in Colorado.

EXOTIC SPECIES

One particularly worrying problem that we humans have caused is the moving of individual animals and plants to quite different parts of the world. We are very familiar with some of the devastation of local species that this has led to. Sometimes it has been accidental, as when rats and cats were left behind by trading boats. Or when domestic pigs escaped and adapted to life in the wild. But often it has been deliberate. Sometimes people wanted to take a piece of "home" when they emigrated to foreign parts—that is how the European house sparrow first arrived in North America. More often, new species were introduced to try to solve a problem—as when mongooses were introduced into the Virgin Islands to kill off snakes. They quickly colonized the area and preyed on the eggs of ground-nesting birds with devastating results. Foxes were introduced into Australia so men could dress up in red coats and hunt them on horses, with dogs. The foxes are now seriously endangering populations of the smaller varieties of marsupials.

Island ecosystems are very vulnerable to the presence of exotic species. On the California Channel Islands, the introduction of feral pigs allowed golden eagles to colonize these islands because piglets are an abundant food source. But the eagles also preyed on the island fox and brought this endemic carnivore to the brink of extinction. As a result of the decline of the foxes, populations of the island skunks increased. The exotic pigs led to the reorganizing of the island's food web and a native predator became prey.

Predator fish were introduced into some African lakes to reduce the numbers of smaller predators that were adversely affecting fish populations whose numbers had been reduced by overfishing. The big predators ate all the smaller predators, then started on everything else, then starved. Indian house crows were imported to Zanzibar to clear up after huge outdoor banquets. Quickly they became a nuisance, swooping down on guests as they ate. Then the authorities tried to get rid of them. Not only was this impossible, but some of these birds managed to make the forty-five-mile journey to mainland Tanzania. During the 1970s I first noticed this new bird, looking much like a jackdaw, and wondered what it could be. Suddenly their numbers escalated with terrifying speed. Within a few years, populations of local birds had been drastically reduced along the coast. The black-and-white pied crows that had greeted the sun with harsh calls each morning for hundreds of years could not compete with these smaller, sleeker strangers. Now they have gone, along with almost all the other native species, whose eggs and chicks have been ruthlessly devoured. The government placed a bounty on the heads of the Indian invaders and crow traps appeared in more and more backyards. But, if anything, their numbers have increased.

When I was in Buffalo, New York, I was told that the plant, purple loosestrife, has taken over roadside ditches, streams, and fragile wetlands. Because it has a beautiful flower, it is also sold as a garden plant and in suitably wet conditions can overtake a garden. Garden shops continue to sell it because it is so popular, despite the fact that it is choking out vast areas of indigenous flora—and it is still spreading. There are literally hundreds of similar stories in almost all parts of the world.

INTERVENTION—OR HELPING?

Often during our long-term study at Gombe, which began in 1960, we have cured, or tried to cure, sick chimpanzees and baboons. We routinely administered antibiotics when the chimpanzees or baboons whom we were studying developed bad infections, and when it seemed necessary, individuals were given medication for deworming. In 1966 an epidemic of what was almost certainly polio broke out among the chimpanzees in the park. Once we realized what was going on, we acquired some oral vaccine and started all of our chimpanzees on a three-week course, putting drops into bananas. Another time the adult female Gilka developed a strange fungus infection that caused her nose, brow ridge, and eyelids to swell grotesquely. We anaesthetized her, performed a biopsy, and then administered medication that helped to reduce the horrible swelling. When Goblin developed two scrotal abscesses, a veterinarian darted him, lanced the wounds, and administered a massive dose of antibiotics. Loretta was darted too when we found her with a poacher's snare pulled cruelly around one hand. Her hand was so gangrenous it was necessary to amputate. When the baboons of Gombe became infected with a syphilis-like venereal disease, we were advised that only massive doses of penicillin would cure the condition. In an incredible veterinary operation, the baboons of three troops (about 150 individuals) were anaesthetized and treated—with no loss of life.

At almost every lecture I am asked a question about interventions of this sort. There are those who feel that we should not "interfere with nature." But we have already interfered greatly with nature, and many of the problems we are treating would not have occurred if the chimps and baboons were not living so close to humans. We can make the argument that every chimpanzee is precious, from a purely genetic point of view, because there are only some 120 chimpanzees within the park and they are now isolated from remnant groups outside. In the old days chimpanzee females moved into the Gombe communities from outside the park. Today the tiny, thirty-square-mile Gombe National Park is surrounded by farmland from which all the forests—and the chimpanzees—have gone.

Resident veterinarians are attached to many of the sites where the great apes are being studied including, now, at Gombe. Chimpanzees, bonobos, and gorillas are all susceptible to human infectious diseases. They are all endangered, and every individual is precious in terms of a diminishing gene pool. But it was not so in the early 1960s, when I began my observations. Chimp habitat stretched right along the shores of Lake Tanganyika and inland east of the rift escarpment. Yet even then I tried to treat the chimpanzees whenever possible. On purely humanitarian grounds. Just as I have always tried to help any suffering animal. We humans cause such massive interference, such great pain, to so many millions of animals. The least we can do is to help when we can—provided we are as sure as possible that our help will not lead to problems of another sort, as was the case when wild dogs were anaesthetized to be vaccinated against distemper, which proved so disturbing that it may have compromised their immune systems, leaving them increasingly vulnerable to other diseases.

THE EARTH CHARTER

Many organizations have been formed during the past twenty years or so by groups of individuals who care passionately about the future of our beautiful planet.

The Earth Charter is a good example of an international organization dedicated to supporting a global effort to help us be wise stewards of life on Earth. This charter establishes fundamental principles for sustainable development. According to its Web site, "the drafting of an Earth Charter was part of the unfinished business of the 1992 Rio Earth Summit. In 1994 Maurice Strong, the Secretary General of the Earth Summit and Chairman of the Earth Council, and Mikhail Gorbachev, President of Green Cross International, launched a new Earth Charter initiative with support from the Dutch government. An Earth Charter Commission was formed in 1997 to oversee the project and an Earth Charter Secretariat was established at the Earth Council in Costa Rica. . . . A new phase in the initiative began with the official launching of the Earth Charter at the Peace Palace in The Hague on June 29, 2000. The mission of the initiative going forward is to establish a sound ethical foundation for the emerging global society and to help build a sustainable world based on respect for nature, universal human rights, economic justice, and a culture of peace."

By acting as stewards, by considering how we live and what demands we make on the natural resources of the planet, we can help to make this a better world for all living beings. And we ourselves shall feel better, less guilty. There is a lot that we can do, each in our own way. We must not lose hope, we must not give up. Above all, we must share what we have learned, encourage our friends to help in this most vital task—the stewardship of the Earth.

But time is running out and we must speed up our efforts; otherwise it will be too late.

THE SIXTH TRUST

VALUE AND HELP PRESERVE
THE SOUNDS OF NATURE

The Sixth Trust concerns the immense damage we have already inflicted on the complex web of life on Earth. So many beautiful animal voices have been silenced in so many places. We cannot bring back creatures who have become extinct, but we can work much harder to reduce pollution, to be less destructive, to save all the beauty that is left.

SILENT SPRING

In 1962 Rachel Carson published *Silent Spring*. She predicted that more and more wildlife would vanish from our fields and gardens as a result of pesticides, chemical fertilizers, and other such products. Today the story of DDT and its long-term devastating effect on the environment is well known. Most people are aware, at least to some extent, that insecticides, synthetic chemicals of all kinds, industrial and agricultural waste, and a whole variety of pharmaceutical products are gradually poisoning land, water, and air. But back in the early 1960s Rachel Carson stood almost alone, fighting to be believed. She carried out endless tests and fought an ongoing war with the giants of the pesticide industry, who repeatedly tried to discredit her work as unscientific and sloppy.

And although Carson has been proven right again and again, we are still throwing new synthetic chemicals into the environment with no way of predicting their long-term effect. And it is not only wildlife that is suffering. We in the industrialized nations have some fifty chemicals in our bodies that were not there half a century ago. More and more people living in heavily contaminated areas are complaining of increasing cases of cancer, especially childhood leukemia, and asthma. In some parts of eastern Europe and many places in the developing world it is particularly bad.

Rachel Carson sounded the alarm. And, fortunately, she had the courage of her convictions. But she had a tough battle before she could get her message out. Governments and industries, out to make a quick buck, tried to suppress her findings. And although her book had a huge impact around the world, many people did not want to listen. Nor do they today. Because we want to dominate nature, separate ourselves from all the nasty creepy-crawlies, prevent caterpillars from nibbling our flowers and vegetables, and enjoy our sterile backyard. And so the spring is quieter, and getting quieter all the time.

In the seaside town in England where I grew up, I used to be awakened at four in the morning by the dawn chorus—the feathered choir that greeted the day with loud and glorious songs. So many birds. At its peak you could barely distinguish the voices of the different species. Now it is a pale reminder of its past glory. There are fewer species represented and fewer individual birds. I used to hear, in the evening, the snuffling and scuffling of hedgehogs in the leaves, but they are gone now. The teeming insect life of my childhood garden is now represented only by the strongest survivors. And our house is in a residential area in a seaside town. In the industrial areas throughout the world, the voices of the animals have long since been silenced by the mechanized clamor of machines, the roar of traffic.

In my beloved Africa, as populations grow, as forests are clear-cut and deserts spread, and as animals are killed or driven to smaller and smaller areas, the voices of nature are going, in some places, with terrifying speed. During a journey in Congo-Brazzaville, we stopped the car as we were driving through a forest. I wandered off by myself, moving deeper and deeper under giant trees, stopping when I could no longer hear the sound of my companions' voices. But I found that I was uneasy. What was wrong, I wondered? And then I realized. There was a silence. I could hear no birds, no rustling as they searched for food in the dead leaves, no wings whirring though branches. There was no bark of a bushbuck or calling from monkey troops. Only the voices of insects broke the utter silence of that wind-still day. Everything—everything had been hunted and eaten.

I longed to be in Gombe, sitting up on my special Peak early in the morning, listening to the morning pant-hoot calls of the chimpanzees, rising in crescendo as they announce their presence to the world, each with his or her own individual voice. And then, often, these calls are echoed by the chorus of the turacos, big dark-crested birds with brilliant scarlet under their wings and bright yellow circles around their eyes. There is the barking and grunting of a baboon troop, and the shrill birdlike calls of a troop of red-tail monkeys or red colobus. Best of all I love to be there in the evening, to hear the singing pant hoot of the chimpanzees, more like music than any other sound they make. One group starts, each chimp calling from his or her nest at the end of the day, bellies full, content with life. A response from over the valley to the south. And more singing calls high up the mountain to the east. A forest paradise. Thinking of it, as I sadly made my way back to the car, there were tears in my eyes.

There used to be elephants in that part of the Congo, but they had long since been hunted to extinction. The biologist Katy Payne was the first to discover that elephants have their own subsonic language, lower than human hearing, whalelike voices that enable the herds to communicate across great distances. And they have their wild trumpeting too—for me one of the most exciting sounds of the African bush along with the majestic, chilling roar of the lion, which gradually fades away in a series of decreasing grunts until all is silent again, everything else quiet, as though listening. Then comes the hysterical giggling or the low whooping cries of the spotted hyena. But now, across Africa, these voices are vanishing one by one.

It is not only on land that we have silenced the sounds of nature. Katy Payne, who studied whales with her husband, Roger, wrote that the world "is losing deep voices—and some of the richest sounds that have ever been heard are in the songs of whales." The singing of the humpback is indeed one of the most beautiful and haunting sounds I have ever heard—and I have only listened to the recorded singing. Thanks to those who have been studying these amazing creatures, we know that the humpback's song is a true song, with ordered, repetitive sequences, like that of a songbird. And the whales are continually changing their songs, reinventing the notes. Most amazing, when the humpbacks of Hawaii change the patterns of their songs, other humpbacks as far away as the Azores will make exactly the same adaptations.

POISONING EARTH

In 1947, when Thor Heyerdahl took his raft, the *Kon Tiki*, from South America to Polynesia, he was shocked to find signs of human pollution—a few bits of oil—in the remotest part of the Pacific Ocean. Alas, there is massive pollution today in all the Seven Seas. We are poisoning the fish and the plankton that the whales and dolphins feed on. Levels of industrial pollution are so high in the St. Lawrence Seaway that the Quebec authorities have labeled all belugas as "toxic," so that when they die and are washed ashore they must be buried as hazardous waste. This is the result of heavy metals, PCBs (polychlorinated biphenyls), and other pollutants having been discharged into the St. Lawrence River for decades from factories run by General Motors, Reynolds Metal, and ALCOA. The PCBs cause bladder cancer, herpeslike dermatitis, gastric ulcers, and tumors in belugas—50 percent of their calves die as a result of toxins in the milk. These wonderful mothers are feeding poison to their much-loved calves. And in their grief they carry the little dead bodies around until they rot away.

Orcas, off the coasts of Washington State and British Columbia, are even more contaminated, saturated with toxic PCBs from the water. There has been a ban on the use of this chemical in Canada and the United States since the 1970s, but the sea is still contaminated, and PCBs are still pouring into the world's waters from industrializing nations. A female orca, tested in May 2002, showed levels of PCBs four times higher than any marine mammal ever tested. Again, it is not only the animals who are suffering. Inuit women were told not to nurse their babies because their breast milk is contaminated. And the fish in the St. Lawrence River are so toxic that pregnant women and children cannot eat them. (I cannot imagine, in that case, they can be very good for men, either!) Everywhere in the industrialized places of the world it is the same.

Recovery from the well-known 1989 oil spill in Alaska's Prince William Sound is slow. Of seventeen seabird species that were affected, only four show very weak recovery from the spill, and nine—cormorants, various gulls, grebes, terns, and murres—show no evidence of recovery. And researchers disagree, some claiming that only one species is recovering. Oil is still harming wildlife—sea otters suffer from liver damage and the guts of harlequin ducks are filled with hydrocarbons. Yet Exxon, who was to blame for the spill, claims, "The environment in Prince William Sound is healthy, robust, and thriving."

THE EFFECTS OF
LOW-FREQUENCY SONAR

There is yet another threat to marine life. In 1998 the U.S. Navy conducted the first of a series of low-frequency sonar tests targeting the humpback whales during their calving season off the coast of Hawaii. Dr. Marsha Green, who has spent her life studying whales, had this to say: "The Cold War is over in our terrestrial world, but in the oceans it is just beginning. . . . The sonar the Navy plans to use—up to 235 decibels—in oceans around the world produces a pressure wave that is 100 billion times more powerful than a loud motorboat engine. This level of pressure can deafen and kill not just whales but other marine life." Both humans and whales are vulnerable not only to the loudness of any sound, but also to its energy, pressure, and vibration. Underwater, the low-frequency sonar can rupture whale and human tissue and cause internal bleeding. Dr. Lindy Weilgart has studied whale communication for seventeen years. She believes these tests could alter hearing abilities in whales and other marine animals over a fifteen-mile radius, and could cause profound behavioral disturbance over an even great area.

In March 2000 sixteen whales of various species beached off the islands of the Bahamas—at a time when the navy was conducting more of its sonar tests. Postmortem examinations found evidence of blood in eyes and brains, and lesions from imploded and exploded lungs. For the first time there was a great deal of critical media coverage, and some of the most influential animal protection societies wrote letters of protest to the navy and to the National Marine Fisheries Service, that had issued the permits, stating that the testing was in direct violation of the Marine Mammal Protection and Endangered Species Acts. At first, the navy dismissed these protests, made a short investigation, and then announced that the mass stranding was "a coincidence." It continued its plans, amid growing international outrage, to test off the coasts of New Jersey and in the Mediterranean. In December 2001 the navy admitted that its sonar tests were the most plausible explanation for the whales' death.

Even where there is no sonar testing, the glorious underwater music is increasingly drowned out by the millions of boat engines that ply the seas, for commerce, pleasure, and military maneuvers. In the cacophony of sound, the singing of the whales—those that remain—is like the exquisite music of a string quartet in the middle of a hundred pneumatic drills.

THE FRAGILITY OF NATURE'S WEBS

One of the most frightening scenarios was written up in the *New York Times* in January 1999. The heading read, "A Mysterious Tear in the Web of Life." It reported that overfishing, along with climatic changes, has led to a crash in the herring and pollack populations. This, in turn, has led to a steep decline in the populations of sea lions and seals who feed on them. Sea lions and seals are an important prey species of orcas, so now the orcas had to hunt other prey, including sea otters, in much shallower water than they usually frequented. This created a dramatic 90 percent decline in the numbers of sea otters. In some places they vanished altogether.

The kelp forests, home to the sea otters, were now affected by a sudden massive increase in sea urchins, a favorite food of the sea otter and the main predator of the kelp. The devastation of the kelp forests affected the multitude of marine life that they normally support. Mussels, fish, ducks, gulls, and bald eagles were all affected, as were the already diminished populations of sea otters, sea lions, and orcas. In addition to that chain of disaster, marine life is struggling also with industrial pollution and military sonar testing.

Kevin Crooks and Michael Soulé have another disturbing story to tell about complex interrelationships, this time involving predators—coyotes, domestic cats, opossum, and raccoons—and their scrub-bird prey, including California quail, wren tits, spotted towhees, Bewick's wrens, California thrashers, greater roadrunners, cactus wrens, and California gnatcatchers, living near San Diego, California. Their research is an example of the importance of long-term projects that investigate complex webs of nature that are not obvious at first glance. Crooks and Soulé found that scrub-bird diversity, the number of different species present, was higher in areas where coyotes were either present or more abundant. Coyotes kill domestic cats where they cohabit. Domestic cats, opossum, and raccoons avoided coyotes by avoiding areas where coyotes were most active. Interactions between coyotes, cats, and birds probably had the strongest impact on the decline and extinction of scrub-breeding birds.

Unlike wild predators, domestic cats are recreational hunters; they kill birds even though they are not hungry. Thirty-two percent of residents bordering the San Diego area where Crooks and Soulé conducted their study owned cats. Eighty-four percent of outdoor cats brought back kills to their homes. Cat owners reported that each outdoor cat who hunted returned on average with twenty-four rodents, fifteen birds, and seventeen lizards to the residence each year—a large number of victims.

Crooks and Soulé discovered that the level of predation on birds appeared to be unsustainable and that many species could not survive being taken in such large numbers. Even a modest increase in predation may quickly drive native prey species, especially rare ones, to extinction. Extinctions of scrub-breeding birds are frequent and rapid. At least seventy-five local extinctions may have occurred in these areas over the past century.

All in all, the disappearance of the dominant carnivore, the coyote, resulted in elevated numbers and activity of predators on birds. The structure of the ecological community in which all of these animals live is also influenced by development, which creates pockets of land between which travel is difficult or impossible.

And there are many other examples. In the United States, wetland areas are hosts to almost one-half of all endangered species; however, they are being lost at alarming rates. About ninety thousand acres of nonfederal wetlands disappeared between 1982 and 1992. Fortunately, a recent survey conducted by the University of Denver showed that 64 percent of the respondents said it was "important" to protect wetlands and 27 percent said it was "somewhat important."

We have shared a few sinister examples of the devastating results that can occur when one part of the intricate web of life is disrupted. And we are interrupting it, in hundreds of ways, throughout the world. Brian Vincent, an eloquent spokesman for conservation, has described our human impact on nature, the terrifying rate at which animal species are becoming extinct, in graphic prose: "Imagine listening to a great symphony . . . the clarion call of the trumpets, the thrilling song of the cellos, the lush velvet melodies of the violin, the euphonious voices of the oboe and clarinet, and the inexorable, dull thud of the timpani. Now imagine if the first violins were silenced. Then the violas were hushed. The French horns quieted. The flutes stilled. Until all that remained was the dull thud, thud, thud of the timpani. That is what humans are doing to Nature's grand symphony. . . . We are silencing the songs of the wild, chord by chord, instrument by instrument, note by note until all that will remain is a mere echo of what was once the music of life."

Is anything to be done? Yes, indeed! Nature is amazingly forgiving, amazingly resilient. If we give her time and perhaps a helping hand, areas that we devastated can once again become clean and beautiful.

Rivers and lakes that we utterly contaminated can be cleaned. It is expensive, but it can—and surely must—be done. For a while, Lake Erie was so polluted with oil that it was a fire hazard. The Cuyahoga River was actually on fire for several days. A river on fire! But these bodies of water, like the Potomac River in Washington, D.C., the River Thames in England, and many more, have been purged of much of their poison. There are fish in them that can be safely eaten. Desert-like areas, scarred by soil erosion caused by clear-cutting, can be restored. Trees, when properly nurtured, will grow and once again there will be shade, relief from the burning sun. Even land contaminated by nuclear waste can be made safe—the technology exists, but governments are reluctant to spend the necessary money.

Coral reefs are considered by many scientists to be the most imperiled ecosystem on Earth. In addition to numerous environmental variables (changes in water temperature and chemistry) that destroy coral reefs, overfishing and destructive fishing—the use of trawls that drag along the ocean floor, underwater explosives to stun fish, and poisons—also destroy coral reefs. In the northeastern Atlantic Ocean, forty-five-hundred-year-old reefs suffer gouges as long as two and a half miles as a result of deep-sea fishing. However, coral reefs often spring back into their full glory much more quickly than anyone had thought possible if they are given adequate protection.

CAPTIVE BREEDING PROGRAMS

Animal species on the very brink of extinction can, with protection in the wild and, perhaps, captive breeding efforts, make an amazing comeback. The peregrine falcon was highly endangered in North America until a few years ago. But captive breeding efforts have been successful, and now this beautiful bird of prey can be seen slicing through the skies in many places where it had become locally extinct.

During 2000 I visited one of the two captive breeding programs, established in 1987, for the California condor. At that time there were only seventeen of these magnificent birds remaining in the wild. A handful of biologists were determined to save the species. In a controversial program—it was considered by many to be a colossal waste of money since most people thought it was bound to fail—the remaining condors were trapped and two breeding groups established. Few people are permitted to visit the breeding condors at the Los Angeles and San Diego Zoos in California. I felt immensely privileged as I looked, through one-way glass, at one of the females in her flight cage. And I had a sensation of awe as I watched her spread her wings to their amazing eight-foot span. What an incredible sight. And what a success, after all. For today there are forty-six released condors flying free in four separate release sites in California and Arizona. The shadows of those magnificent eight-foot wing spans move once again across the ancient homelands of the California condors—thanks to that small but determined band of stewards. And now flying with those bred in captivity, the first two of the original seventeen have been set free to share, with the younger birds, the ancient lore of the wild condor. As of February 1, 2001, there were a total of 161 condors, 115 still in captivity. In February 2002, a condor egg was found in southern California. It was only the third confirmed egg laying by California condors who were released into the wild after being raised in captivity. And in April 2002 the first condor chick to be conceived and hatched in the wild was observed. These are first steps toward the goal of increasing condor numbers through breeding in the wild.

Here are some more heartwarming success stories. DDT, which was used to eradicate malaria and for agricultural pest control, was partly responsible for the near elimination of the Mauritius kestrel. Along with predation and the decimation of its habitat, the number of Mauritius kestrels declined drastically to two known breeding pairs in 1973. Successful recovery efforts, which included taking kestrel eggs from nests in the wild, allowing eggs to be incubated in captivity, rearing chicks in captivity, and then returning them to the wild when they were about two weeks old to be reared as adopted young by a wild breeding pair, were undertaken by scientists working with the Gerald Durrell Endemic Wildlife Sanctuary on Mauritius. As a result of this and other efforts, the kestrel population was doubled each year. There are about eight hundred Mauritius kestrels currently alive.

Twenty-five years ago there were only thirteen Taiwanese Seika deer and they were all in various zoos in Southeast Asia. Again, captive breeding worked. There are now two herds, each of over thirty individuals, in two very large national parks in Taiwan. And many other animal species are also struggling to make a comeback. This is why I have hope. As more and more people learn about the damage we have inflicted and about the amazing miracle of life on earth, so there is a growing band of concerned and caring people, eager to help in the struggle to save what is left.

One incident, described by Brenda Peterson in *Build Me an Ark,* cries out to be mentioned. It took place at a meeting that was called to try to convince local farmers to support the plan to reintroduce wolves into the Northwest. Two wolves from a wolf rescue program were taken to be ambassadors. They were captive-born, but not domesticated. One of the ranchers was particularly concerned about her livestock if wolves returned and seemed reluctant to believe that she would be compensated for any losses. Indeed, she was strongly against the idea.

Merlin, the big black male, was the first to approach the people present, who were sitting on the floor. He paused to sniff at some and ignored others. Suddenly he went over to the ranch woman and stood, staring into her eyes. Then, unexpectedly, he lay down beside her for a while. She sat motionless as he looked into her eyes. "He's . . . he's really something," she said. "He does have a way of getting right up into your heart, doesn't he?" For minutes he lay there and then he sat up, sniffed her, and finally, briefly, pressed his forehead against hers.

We cannot care deeply if we are fearful of or have no knowledge of that which we are asked to save, and if we do not care, we shall not help. And it is only when enough of us help that we can ensure that wolf music will endure into the future so that our great-grandchildren can stand still in the moonlight and hear that most thrilling sound, inextricably linked to our perception of the wilderness—the howling of the wolves. And all the other sounds of nature, as varied and diverse as the amazing creatures who produce them: the song of the gibbons, the long call of the male orangutan, the eerie call of the howler monkey, the heart-stirring calling of the coyotes and jackals and the whooping of the hyenas, the songs of the avian choristers, and the haunting underwater symphony of the whales and dolphins. And all the others, the grunts, squeaks, chips, barks, meows, along with the innumerable insect voices around the world. And the rustling of the wind in the leaves of the forests. The sounds and songs of nature. Can future generations ever forgive us if we allow these voices to be stilled forever, preserved only on tape?

THE SEVENTH TRUST

REFRAIN FROM HARMING LIFE
IN ORDER TO LEARN ABOUT IT

In the Seventh Trust, we discuss various ways of learning about the lives of animals and the places where they live. Humans are intensely curious, and this curiosity, though it may provide us with fascinating information about animals, can also be extremely invasive and even destructive. We stress the need for restraint and consideration for the creatures and habitats we study.

LEARNING ABOUT ANIMALS WITH CARE

We can learn about living beings in a number of ways. We can read about a species in books and ask questions. We can quietly observe, write down what we see, and analyze the data—which usually raises more questions, so that we must return to our observations. Or we can proceed along the path to knowledge in an ever more invasive way.

The Nobel-laureate ethologist Niko Tinbergen learned about the nesting behavior of gulls by substituting eggs of different sizes in their nests and observing their behavior. Konrad Lorenz, who shared the Nobel prize with Tinbergen and Karl von Frisch (famous for his studies of bee language), brought birds into his home from the wild and studied their behavior in cages. Charles Carpenter observed gibbons in the wild, then shot the family group to find out the age and sex of the individuals. Many biologists use capture guns to anaesthetize their study animals so that they can fit them with radio collars to track their movements or telemetry devices to measure heart rate and other physiological data. Biology students are taught about living systems by dissecting dead animals. And, in the laboratories of scientists seeking answers to a whole variety of questions—some of which are relevant to lines of inquiry designed to improve (or, in the case of biological and chemical warfare research, to destroy) human health, and some of which are primarily to acquire knowledge for its own sake—animals of all sorts are subjected to a variety of horrifying research procedures, many of which are profoundly cruel.

EDUCATION

The attitude of children toward animals is, as we have seen, shaped by their early experience and the example of key figures in their lives. At school, children learn from books, films, and, increasingly, from the Internet. Those who are fortunate go on field trips so that they can see animals in their natural habitat. Sometimes animals are taken into schools. A class may look after some small creature. All of these methods of learning can generate a sense of wonder.

But, in addition, biology students learn about life through handling dead animals. Often dissection is repellent to a sensitive child, but every time he or she is forced to cut up a once living creature in class, as we have noted earlier, it will be that much easier the next time. It used to be common to force students to kill the frogs they dissected by pulling out the spinal cord. This was known as "de-pithing." This practice continues in the United States and is common in many other countries. Once a child kills even a simple living thing—such as an earthworm—it will be easier to kill a more complex one. Once sensitive young people will become callous, "all pity choked by custom of fell deed."

Moreover, the children are getting the implicit message that it is acceptable to exploit animals for their education, that exploitation of the weak by the strong is okay. Although some children become even more amazed by the intricacy of organisms through dissection, others lose some of their respect for the miracle of life. And, quite apart from the damaging effect on the students' attitudes toward animals, the provision of animals for educational purposes has become an industry and people are making huge profits. About 170 species, including at least 10 million vertebrates, are killed annually for education in the United States alone. It has been estimated that about 90 percent of the animals used for dissection, including frogs, turtles, and fish, are wild caught.

Can anything be done? As the animal rights movement took hold in America, more and more children felt badly about dissection. One of them, Jennifer Graham, refused to do it. She was told that she could not take her examination in biology, even though she demonstrated a knowledge of biology that was quite as good as that of her peers. There were articles in the local paper, and the Humane Society of the United States (HSUS) decided to take up her cause. They helped her to challenge the school authorities in court—and they won the case. This was a landmark victory and led the way for others to maintain the right not to dissect once living creatures killed for the purpose.

A visit to a natural history museum can be instructive and enjoyable, especially when we know that the stuffed animals died in zoos or were reclaimed from roadkills and not harvested for the exhibits. But I can never forget the first time I went behind the scenes at the natural history museum in Nairobi, Kenya. I had asked the ornithologist a question about a bird, so he took me to see the collection. I could not believe what I was seeing—drawer after drawer of dead, dried, stuffed birds, laid out in rows with their beaks parallel with the bottom of the drawer, their legs stretched out beside their tails. In one drawer were at least twenty specimens of one species of songbird, each one smelling of musty preservative, with small identifying labels tied to the legs. And it was the same in all the drawers of the smaller birds. I closed my eyes. I could imagine the fear of a bird as it became entangled in one of the "mist nets"—those flimsy traps stretched across the flight paths of the forest birds. The increasing panic as it struggled to get free, the pounding of the tiny heart as a dreaded human approached. And then the snapping of the neck vertebra. One more life sacrificed at the altar of science. Somehow I was no longer interested in getting the answer to my question.

HUMANE NONANIMAL ALTERNATIVES

There are many other ways to learn about animal biology—nonanimal alternatives—that are less destructive of life, many of which are not only as instructive as dissection, but are better—and often less costly. Jonathan Balcombe, in his book The Use of Animals in Higher Education, *summarizes the cost of alternatives compared to the use of common dissected animals. For cats, reusable alternative materials (an anatomy model, a dissection video, and a video player) cost about $1,865, whereas the use of cat specimens costs between $5,000 and $8,300. A CD-ROM for commonly used fetal pig dissections costs about $20 compared to a single fetal pig specimen, which costs as much as $24.*

Balcombe also compared the educational effectiveness of alternatives such as computer software and models. Studies show that alternatives often are at least as good, if not better, for achieving intended educational goals. For under-graduate college students, veterinary students, and medical students, equal knowledge or equivalent surgical skills were acquired using alternatives. The educational effectiveness of nonanimal models was not less, and educational quality was not compromised.

For example, in a study of 2,913 first-year biology undergraduates, the examination results of 308 students who studied model rats were the same as those of 2,605 who dissected rats. When the surgical skills of 36 third-year vet-erinary students who trained on soft-tissue organ models were compared to the surgical skills of students who trained on dogs and cats, the performance of each group was the same. In a study of 110 medical students, students rated computer demonstrations higher for learning about cardiovascular physiology than demonstrations using dogs. Richard Samsel and his colleagues, at the Uni-versity of Chicago, found that first-year medical students rated both computer and animal demonstrations highly in teaching cardiovascular physiology, but that the computer-based sessions received higher ratings. Virtual surgery has been shown to be an effective alternative. For example, at the University of Cal-gary in Canada, virtual reality images help to eliminate the use of laboratory animals and cadavers in training and research.

Many medical schools have come to recognize that "hands-on" experience is not needed in certain parts of their curricula. Currently, 90 of 126 American medical schools (71 percent), including such prestigious institutions as Har-vard, Yale, Columbia, Duke, and Stanford, do not use live-animal laboratories to train medical students. Medical students can select nonanimal options or choose not to participate in a particular laboratory that uses animals even if no alternatives are offered in 125 U.S. medical schools, the sole exception being the Uniformed Services University of Health Sciences. Similar trends are devel-oping in veterinary schools, in which terminal surgery laboratories have been dropped from the curriculum. In a survey taken in 2001 to which 40 percent of the 85 accredited veterinary schools responded, almost one-half had no inva-sive or terminal procedures in their curriculum.

There are so many ways in which we harm animals as we try to learn about them, even if sometimes they are quite unintentional. We field biologists, by our very presence, may influence movement patterns, feeding habits, and even the composition of groups of the animals we are studying.

In a long-term study of coyotes that Michael Wells, numerous student researchers, and I conducted in the Grand Teton National Park outside of Jackson, Wyoming, we were very sensitive to the way in which our presence influenced the behavior and well-being of the animals we were observing. We found that shiny cameras and spotting scopes made the animals uneasy, so camera bodies and spotting scopes were painted dull black so they wouldn't reflect light. When visiting dens we wore the same clothes, so that roughly the same odors were present and so that we presented a similar visual image to the coyotes. The coyotes eventually adapted to our presence and showed none of the skittishness that characterized their behavior very early in our study. The data we collected were less tainted by our presence and more truly representative of the natural behavior of those fascinating animals, some of nature's most cunning and tireless tricksters.

There are many ways to learn about animals without harming them. For example, when researchers want to gather information on paternity, answers can be obtained from DNA profiling based on hair samples and feces; there is no reason to handle individuals to draw blood. This method has been used successfully with wild chimpanzees and other animals.

Other methods of data collection are far more disturbing for the animals. Researchers trying to document frequency of egg loss in white-fronted chats visited selected nests on a daily basis. These visits had an adverse effect on the birds, as more eggs were lost to predation and fewer chicks were hatched than in surveys in which nests were visited only once at the end of a typical period of incubation. Population surveys, when researchers gather information on numbers of males and females living in selected areas, movement patterns, and overall population size, can be disturbing. One such survey of Adélie penguins involved not only approaching the colony, but also flying over the birds in

aircraft. The penguins showed profound changes in behavior, including devia-
tion from a direct course back to a nest and increased nest abandonment. The
overall effects included a decrease of 15 percent in the number of birds in a
colony.

Even more disturbing, certainly for some species, is the fitting of radio collars. Individual animals are trapped or shot from a car or helicopter with a capture gun, tranquilized, and fitted with a collar that gives off a unique signal enabling that individual to be tracked from afar. A great deal of extremely useful information has been collected in this way. It is particularly helpful when trying to determine the range and movement patterns of shy or nocturnal animals such as jaguars, cougars, and tigers. But despite the fact that this is usually done by researchers with the well-being of the animals at heart, there are often accidents. In spring 2001 three wolves in Denali National Park in Alaska died after being darted. And the whole procedure, especially when wolves, bears, and other animals are darted from helicopters, is extremely stressful.

One researcher who was studying hyenas at Ngorongoro Crater in Tanzania left radio collars in place at the end of his fieldwork. When Hugo and I began observing hyenas there, a few years later, we found one adult male who had clearly been collared as a youngster—we had to find someone to tranquilize him and remove the collar, which had become so tight that he could hardly swallow. This showed an unacceptable disregard for the well-being of individual animals. Some researchers tranquilize animals to mark them for identification purposes. One scientist fixed colored tags onto the ears of the lions he was studying. I was horrified when I saw them—they were so large and conspicuous (so the scientist could easily identify individuals) that they most probably interfered with the lions' hunting success. Prey animals are quick to notice anything unusual in their field of vision.

It is a disturbing fact that many scientists handle animals, for a variety of reasons and in many settings, in ways that would not be tolerated by nonscientists. All too often procedures are undertaken that, although providing a researcher with material for a Ph.D. thesis, paper, or book, are definitely not in the best interests of the animals. However, as more scientists admit that the animals have feelings and even allow themselves to feel empathy with their subjects, it is hoped they will themselves question whether invasive or stressful methods are really necessary, whether there is not a better way of doing things.

ECOTOURISM

Many people, anxious to learn more about animals firsthand, go out into wilderness areas to try to see, observe, photograph, or film them. In doing so, they can affect the lives of animals absolutely unintentionally. The pollutants from outboard motors can have devastating effects on fish and other creatures of our oceans, lakes, and rivers. And the major noise pollution is a horrible intrusion into the once peaceful underwater world. Manatees, those slow and gentle giants, get the most terrible injuries from collisions with outboard motors. I saw this for myself at a wonderful manatee sanctuary in Florida.

Many people spend their annual vacations traveling to wildlife refuges in different parts of the world, and in many places the animals are so used to people that cars can get quite close without seeming to disturb them. In the great national parks and game reserves of east and southern Africa, visitors can drive all day through herds of plains animals, watch lions or wild dogs hunting, and see the sun set over water holes where elephants drink and play. These tourists have, for years, brought foreign exchange into countries that needed it badly. Various studies have been carried out on the impact of tourism on the land and the animals—and where too many people and too many cars were allowed, researchers found there was considerable disturbance. In particular, driving off the roads was very damaging for some fragile habitats. Tourism in the forests of the chimpanzees, gorillas, and orangutans poses a threat, for they are all susceptible to human infectious diseases, and visitors from overseas can carry bacteria and viruses to which the apes have no natural immunity.

However, while tourism is indeed a two-edged sword, it seems that on the whole, provided the number and behavior of visitors is strictly controlled, the benefits outweigh the costs. Indeed, before the terrible genocide in Rwanda, gorilla tourism was the second highest foreign exchange earner in the country, and many people believe this is what saved gorillas. Thus, even when the armies of both sides were actually in the Virunga National Park, first made famous by Dian Fossey (whose book *Gorillas in the Mist* provides a detailed account of her dedication and long-term research), the gorillas were not harassed, although other wild animals were killed for food. It seems that both Hutu and Tutsi wanted to preserve the gorillas for future gain.

Of course, it is sad to have to put a monetary value on the wilderness and on animal species. But until the wealthy nations can agree to pay an annual "rent" on huge areas of land, it seems likely that governments in the developing world will exploit their natural resources in any way they can—granting concessions for oil, timber, or mining or opening it up for development. After all, it is happening in America, even in national parks created to conserve the natural world. The presence of genuinely interested visitors is surely better than that.

I have spoken to so many people who have told me that their experience in a wilderness area where they could, perhaps for the first time, encounter wild animals face to face was life-changing. They can never think the same again about captive animals. They can never read about the destruction of the natural world without hurting inside because of the suffering of the animals who live there.

I visited Yellowstone National Park for the first time recently with Tom Mangelsen as my guide, and I saw my first wild bears—a grizzly who suddenly stepped into the road as we passed, so close I could have touched him, and a black bear with her two cubs. It was wonderful. And what impressed me so much was the number of people, from different parts of the United States, who went there on vacation. It was cold toward evening. But there they were, with anoraks and binoculars and spotting scopes, checking on the grizzlies, and especially on the reintroduced wolves. Yes, they told us, they were there early every morning and late every evening, when there was the best chance of seeing the bears and wolves. Every day of their holidays. These were not biologists, for the most part. They were just people who cared, who wanted to learn. These are the people who will help to keep the Seventh Trust, who will try to leave light footprints and teach their children to respect all life.

By leaving lighter footprints, we can not only ensure that future generations will be able to enjoy the activities that we enjoy, but also that the animals will behave as the beings they are and not as individuals—artifacts—whose behavior is changed by our presence. There is much pure and exciting joy associated with understanding and appreciating what it is like to be another animal and allowing our animal kin to tell us what life is like for them in their own worlds. The purer the details, the greater the awe and mystery. There is no need to harm life in order to learn about it.

THE EIGHTH TRUST

HAVE THE COURAGE OF OUR CONVICTIONS

We show, in the Eighth Trust, that by steadfastly voicing our concerns, upholding our beliefs, and taking action, we can bring about change. Once a sufficient number of people take action, we can make changes that benefit all animals and the Earth.

WALKING OUR TALK

Each day of life is a test. Do we actually "walk our talk"? Do we practice what we preach? Do we think about the footsteps we leave as we move through life? And, above all, do we have the courage to stand up for what we believe in when others laugh at us? Or when they threaten us? If we care, truly care, about the natural world and about the well-being of animals, then we must be prepared to face the challenges, both small and large, that come our way. We must stand up for our beliefs and, above all, put them into practice. We must not despair, but live with hope for the future.

Much of the destruction and pollution around the world is mindless, due to lack of education and understanding. "Only when we understand, can we care." But a great deal is deliberate. For corporate or personal economic benefit or to gain political power, industry and governments take and support actions knowing full well what the consequences will or could be. As the general public becomes more aware of such plans, there is increasing hope that pressure can be placed on the giants of industry and commerce—and, too, on political will for change. There are three ways in which we can prove, by our words and actions, that we care. First, we can protest by spreading information, taking part in peaceful demonstrations (that will be more difficult after the terrorist attacks of September 11, 2001), and by writing letters. Second, we can refuse to buy products from companies with bad environmental or humanitarian practices, and we can refuse to watch entertainment that involves any form of cruel exploitation. And third, perhaps most important of all, we can live our own individual lives so as to leave the lightest possible footprints—by doing all those little things like saving water and energy, recycling and reusing, and trying not to pollute our world.

Often, before the general public is aware of certain issues, there are those who know what is going on, but do nothing about it. But there are some who will speak out against all odds, who will refuse to be silenced even in spite of death threats. As we have seen, Rachel Carson was threatened with lawsuits by some of the petrochemical giants if she dared publish *Silent Spring*. Knowing that her science was accurate, she went ahead and published it anyway—and its impact was quite extraordinary, causing new investigations into the effect of pesticides on our food, on ground water, and on wildlife throughout the world. Rachel Carson led the way.

Now many of the projects planned by industries in cahoots with governments are being one by one brought to the public's attention. To our attention—yours and mine. Shall we dare stand up and make our voice heard? Shall we protest issues of environmental destruction or cruelty, once we have heard about them? Shall we opt to buy products made by conscientious companies, even if they cost a few more cents? Shall we boycott goods produced with little or no regard for the well-being of the other animals with whom we share this planet, boycott entertainment that harms them or their image? Do we, indeed, have the courage of our convictions?

DRILLING IN ALASKA

As we are writing this book, the future of one of the most unspoiled tracts of land in the United States hangs in the balance. The Arctic National Wildlife Refuge (ANWR) is a glorious area of seventy-seven-thousand square kilometers in northeastern Alaska. The Bush administration would like to drill for oil there. Those who think the term "refuge" implies safe harbor should think again. When it comes to refuges for wildlife, the term means as little as it did to the Hutu militia of Burundi who slaughtered Tutsi women and children who had taken refuge in a church. After all, more than three hundred "wildlife refuges" have recently been opened to hunting in the United States.

When I traveled to Cape Cod (Massachusetts) in March 2002 to teach residents about coyote behavior and to discuss how they could coexist with what some people call "nuisance" coyotes, I visited the Monomoy National Wildlife Refuge. A sign at the entry of the refuge indicated that animals on the refuge are to be protected from human disturbance, but I learned that this is not so. In mid-April 2002 a sharpshooter commissioned by the refuge killed nine coyote pups on the refuge and displaced a tenth pup, who later died in captivity. To date, thirty-four coyotes have been killed on this so-called refuge. Surely, killing animals is a form of human disturbance.

The Arctic refuge is home to many animal species such as polar bears, grizzly bears, Dall sheep, wolves, wolverines, lynx, porcupine, caribou, golden eagles, and thirty species of shorebirds. There has been considerable debate about whether or not to explore six thousand square kilometers of the reserve's coastal plain for petroleum-based products to maintain consumptive lifestyles. Joel Berger, one of the world's leading conservation biologists, analyzed the situation on the ANWR to determine if we should even consider exploiting this pristine area, one of the world's most significant remaining intact ecosystems. He and many other experts concede that we just do not know what effects exploration will have, but there is no doubt that there will be many negative influences on the lives of the animals who live there. Irreparable damage will be done to one of the world's most incredible natural treasures, and this will be a significant loss to all humans. Berger and his colleagues also note that if we continue to ruin intact ecosystems, future researchers will be unable to make comparisons between natural, intact ecosystems and those that humans have trespassed so that we can assess the impact of our intrusions. Essentially, "real" nature will never be gone forever.

Little is known about the intact ANWR ecosystem, and experts caution against tampering with any ecosystem until we have basic information about interrelationships among the various animals who live there and how animal and plant communities are related to one another. Retaining the biological integrity of the ANWR and other ecosystems is the main goal of conservation biologists, even if it means that some of us have to change our lifestyles and use less energy. Some Republican members of Congress went as far as trying to tie drilling in the ANWR to an unrelated bill for increasing the retirement benefits for railroad workers, but the vote failed on December 3, 2001. In April 2002 the United States Senate again rejected a plan to drill in the ANWR. The American environmental community can be credited with halting, at least for the moment, another ill-conceived energy plan. It is hard to believe we can even think about ruining such a pristine habitat.

It is important to realize that fossil fuel is not an infinite resource, and that an increasing number of new technologies offer alternatives to oil. I know many people who are using electric or hybrid cars, vehicles running with fuel cells, and other alternative engines. They may be more expensive to buy initially, but they are cheaper to run. Furthermore, the more people who buy them, the more quickly the price will come down. But how many, among those for whom the price is no object, are actually prepared to buy such a vehicle? So far, tragically, very few. In addition to these cars, there are countless ways in which we, as individuals, can economize on our use of fossil fuels and other forms of energy, some of which are very simple.

After the attacks on the World Trade Center and the Pentagon, a war against terrorism was launched. This has provided a platform for those who wish to exploit places such as the Arctic refuge in Alaska—the fuel is needed for the war effort. Therefore it will be seen as unpatriotic to protest exploring for oil in the wilderness. Indeed, many Americans are afraid to show concern for environmental or animal issues in the face of human suffering. I believe this is absolutely wrong—I think it is more important than ever before to continue to work for these issues. If we allow the fragile health of our planet to be further destroyed, we shall, ultimately, be handing a new victory to the terrorists—for it will damage, perhaps permanently, the quality of life for our grandchildren and theirs.

There is another way of looking at this whole issue: as people and machines move into a wilderness area, such as the Arctic refuge, the homelands of the animals living there are destroyed, many are killed, and hundreds of thousands are terrorized. Is not this, from the animals' point of view, the equivalent of a terrorist attack? I find this profoundly disturbing. It applies to so much human activity such as logging, mining, development, road construction, the building of dams—the list is endless. It was Mahatma Gandhi who wondered why, when someone harms a man-made object such as a building or work of art, it is described as "vandalism," but when someone harms an object created by God, it is simply shrugged off as "progress."

GENETICALLY MODIFIED
FOOD AND DRUGS

One of the biggest assaults on natural processes involves the burgeoning indus-
try that produces genetically modified food, or "Frankenfood." Japanese scien-
tists have even bred pigs with spinach genes and claim they have produced
healthier pork. We know very little about the possible threats to human health
posed by these foods, and we do not know the effects of genetic engineering on
natural processes. Genetically engineered crops are created when scientists take
genes from one organism and insert them into another organism. The inserted
genes can come from an entirely different species than the host organism. Thus
scientists may insert a gene from an animal or a bacterium into a vegetable. In
many cases, insertion of the gene allows a plant to become resistant to a specific
herbicide (agent used to kill weeds) or to produce its own internal pesticide.

The USDA, which promotes genetically engineered crops, claims they pro-
duce higher yields and are needed to feed an exploding world population. The
USDA claims that genetically modified crops need less pesticide and herbicide
protection, meaning fewer chemicals will be sprayed into the environment. But
critics worry that there is been no long-term testing of the technology. They also
are concerned that insects will become resistant to plant-produced pesticides
and that genetic engineering may cause unexpected mutations that produce tox-
ins or allergens dangerous to human. Once introduced into the environment, it
is impossible to contain or recall them; the effects are essentially irreversible.

Even as the safety of genetically engineered food continues to be hotly
debated by scientists, governments, and consumers around the world, they are
gaining wide acceptance among producers. Less than 10 percent of soybeans
and corn planted in the United States during 1996 was genetically engineered.
Within two years, about 33 percent of corn and 40 percent of soybeans were
genetically modified types. Worldwide there are about a hundred million acres
of various transgenic crops, including cotton and soybeans. Some other geneti-
cally modified organisms include corn, canola, flax, papaya, potatoes, tomatoes,

peppers, squash, radicchio, enzymes, dairy products, and pet food. Soybeans are the source of soy flour, soy oil, lecithin, and soy protein isolates and concentrates. Soy and its derivatives are found in about 60 percent of processed foods, including breads, candies, cereals, chips, chocolates, cookies, crackers, enriched flours and pastas, fried foods, frozen yogurt, ice cream, infant formula, margarine, protein powder, sauces, soy cheeses, soy sauce, tamari, tofu dogs, veggie burgers, sausages, shampoo, bubble bath, cosmetics, and vitamin E. Transgenic biopharmaceuticals include antithrombin III, angiotensin, insulin, and prolactin.

Genetically modified foods, I believe, are one of the more frightening technologies that we are carrying into the twenty-first century. It really seems there is something sinister afoot. So often an item creeps into the popular press that is bad news for giant multinationals, such as Monsanto, and then simply vanishes. In the United Kingdom, for example, a Polish scientist working for the government published results of his investigation into the genetically altered Monsanto potato. He fed it to rats who, he reported, suffered from a variety of physiological problems including shrinking of brain tissues and damage to other internal organs. A couple of days later he appeared on television admitting that his experiments had been faulty and his data erroneous. This was followed by his resignation.

About a year later I happened to tune into the BBC World Service just in time to hear a news item on the same scientist reporting that a number of other scientists in different European countries had replicated his potato and rat research. Since then I have heard nothing. Soon after that story, another short report appeared in the *Daily Telegraph* in the UK. This was about test fields of genetically modified rape seed in the south of France. When these plants bloomed they were visited by hundreds of bees, and by the hundreds the bees died. Biologists who investigated the deaths (since it affected the beekeeping industry) concluded that something, probably the pollen, they thought, had damaged the bees' central nervous systems so that they could not find their way home.

Even more frightening, bees sometimes fly as far as two miles from the hive during foraging. Carrying pollen from modified plants, they could cross-fertilize "normal" crops, and organic farms with test fields within range may lose their certified status. There is evidence suggesting that caterpillars of the Monarch butterfly died after eating leaves that had been contaminated with pollen from genetically modified corn. In 2001 a number of people developed alarming allergies apparently as a result of eating genetically modified corn that had been developed for animal feed. What will be the long-term cumulative result of feeding animals with such food, both for the animals themselves and for the humans who eat them?

There is also a problem with genetically engineered drugs such as Humulin (human insulin). It has been reported that clinical trials were never completed and that in 1995 Humulin was the eighth most reported drug causing adverse effects in the United States. In Canada, different brands of rDNA insulin have been responsible for more than 450 reports of adverse drug reactions, including eight deaths associated with the use of synthetic insulin.

Cloning has been used to produce, according to some researchers, "super" animals (and even "pet replacements," despite the fact that there are millions of animals in shelters waiting to be adopted). But results from genetically engineered cloning experiments are extremely mixed and controversial, scientists cannot agree about how successful cloning has been or will be, and the perils of misuse are both daunting and haunting. Dolly, the first cloned sheep, showed premature aging, perhaps because her cells came from an old sheep, but six cloned heifers showed signs of being younger than their actual ages. Is there a fountain of youth in the offing and is this really good?

Japanese researchers discovered that cloned mice often die early in life due to liver failure and pneumonia. An attempt to clone an endangered wild ox, the gaur, failed when the baby gaur, Noah, died of a bacterial intestinal infection forty-eight hours after birth. Plans to clone giant pandas are in the offing, as are plans to clone humans. But to what end? Is cloning the panacea for which we have been waiting, the elixir that will make the world a better place, or is it just another example of science run amok? Surely we do not know enough about the long-term effects of cloning for researchers to wantonly try to clone this and that.

So is there anything that we can do about all this? Indeed, there is. We are the consumers, and we can refuse to buy products we believe have been produced in an irresponsible way. In many European countries consumers are winning the war against genetically modified foods. In the United Kingdom, stores must list any genetically modified ingredients on the labels of all products (although sometimes this is almost impossible). Farmers can no longer grow genetically modified crops. One whole shipment of seed, after planting, was found to have been mixed with genetically modified seeds; the farmers were ordered to plough up the fields and were compensated by the government. Most scientific trials outside the confined spaces of the laboratories are banned. All of this happened after protesters, in France and Germany as well as the UK, repeatedly pulled up genetically modified crops from test fields around the country. These protesters were willing to go to jail for their strong belief that genetically modified food was dangerous. Whole fields of genetically modified rice, grown from so-called terminator seed (because the plants it produces cannot themselves produce fertile seeds) were ripped out of the ground by demonstrators in India. Japan is adamant that it will not introduce genetically modified foods. Public pressure works.

In the United States the public is at last becoming aware of the dangers of genetically modified food. Hershey Foods and M&M Mars have asked growers to avoid planting transgenic sugar beets, not because they question (publicly anyway) the technology, but rather because many consumers reject food with such ingredients. Indeed, in response to public concern, the USDA has spent millions of dollars to buy up genetically altered corn seeds to keep them from getting into the food chain.

DRUGS, ANIMAL SUFFERING, AND HUMANE ALTERNATIVES

Women could prevent a great deal of equine suffering if they refused to buy Premarin, an estrogen replacement used worldwide that contains the urine of pregnant mares. Premarin is used in estrogen therapy to reduce symptoms of menopause and reduce the chance of heart disease, and is prescribed also for women who have had a hysterectomy to eliminate their risk of osteoporosis (brittle bones). However, estrogen therapy has been shown to increase the risk of breast cancer by up to 70 percent after nine years of use, and chances of endometrial cancer were suggested to be 5.6 times greater when estrogen use exceeded seven years.

Some of the facts surrounding the production of Premarin are not well known. For six to seven months during pregnancy, mares are kept in tiny stalls in which their movements are extremely restricted. They suffer from stiff joints and lower limb abnormalities from having to stand constantly with little or no exercise. The mares must wear rubber urine collection bags at all times, which prevents them from lying down comfortably. They are not given much water because the more concentrated their urine, the more it is worth. Mares with oozing sores are not treated with antibiotics because it would contaminate their urine. About seventy thousand foals are born on Premarin farms, of whom about thirty thousand replace their mothers and forty thousand are sold to food factories and slaughtered for horse meat. Foals nurse and remain with their mothers for three or four months rather than the normal six-month period. In excerpts from a farm investigator's daily notes, one farmer said, "Sometimes a mare will get ornery and kick or nip at you because they stand for so many months. . . . We had to pop them on the nose to keep them in line when that happened."

To prevent the ongoing mistreatment of mares used for Premarin, you can choose an alternative synthetic drug such as Estradoil, Estropipate, or Estrone. In California, due to the hard work of Cathleen Doyle, of the California Equine Council and Save the Horse Campaign, Proposition 6, landmark legislation banning horse slaughter in the state, was passed. Premarin farms are no longer allowed in California. In Peachland, Canada, Ray Kellosalmi, a physician and surgeon, and his wife, Noreen Nawrocki, run a sanctuary for foals whose lives have been ruined in the Premarin industry. They spend about ten thousand dollars a year and receive no free services from veterinarians, farriers, or food suppliers. They grow their own hay and apples.

Millions of animals—dogs cats, rats, mice, guinea pigs, and rabbits—are used for testing drugs as well as nonessential cosmetic products such as deodorants, shampoos, soaps, and eye makeup. Lethal-dose tests try to measure the toxicity or potential harm of products using live animals. Animals receive a single dose of the substance to be tested either orally, intravenously, by stomach tube, by inhaling a vapor powder or spray, or by having it applied to the skin. The dose at which 50 percent of the animals die is called the Lethal Dose 50 (LD50) and the dose at which 100 percent die is called LD100. Many animals become sick and suffer greatly, experiencing convulsions, seizures, muscle cramps, abdominal pain, paralysis, and bleeding from the ears, eyes, nose, and rectum. If too many or too few animals die, the tests need to be repeated.

In addition to the inhumanity of these tests, results are only specific to the condition in which they were used, so they cannot be generalized from species to species or even between males and females of the same species. The LD50 is often used to estimate the safe dose of a given product for humans. For example, paraquat was introduced as an herbicide in 1960. Because the LD50 for rats was 120 milligrams per kilogram of body weight, it was thought that humans exposed to fewer milligrams of paraquat would be safe. However, in twelve years more than four hundred humans died from exposure to this chemical, and it was estimated that the LD50 was much smaller, about 4 milligrams per kilogram of body weight. Furthermore, in the United States more than one hundred thousand people die annually from side effects of animal-tested drugs; one person every fourteen seconds is hospitalized by them; and one in seven hospital patients is there because of the ill effects of animal-tested drugs. Ultimately, many products cannot be marketed because these tests are invalid, but animals suffer nonetheless.

THE DRAIZE TEST ON RABBITS' EYES

Rabbits are wonderful animals. Anyone who's lived with one of these furry creatures knows how loving they can be and how much fun they are to watch as they go about their daily rituals. For a long time, rabbits have been used to test cosmetics.

The Draize test for eye irritations is named for a Federal Drug Administration (FDA) scientist, John Draize, who standardized the scoring system of a preexisting test for eye irritation in 1944. In the Draize test, a liquid or solid substance is placed in one eye of each subject in a group of rabbits. The changes in the cornea, conjunctiva, and iris are then observed and scored. The rabbits' eyes are inspected at twenty-four, forty-eight, and seventy-two hours and at four and seven days for injuries and potential for recovery. The Draize test is very painful, and the rabbits experience great suffering. Consumer protests against widespread use of the Draize test created the momentum that led to the development of nonanimal alternatives to many types of animal testing. Animal activist Henry Spira's campaign against the Draize test unleashed a growing movement against causing animals discomfort. By 1981 the cosmetics industry awarded $1 million to the Johns Hopkins School of Hygiene and Public Health to establish the Center for Alternatives to Animal Testing (CAAT).

TESTING ENVIRONMENTAL POLLUTANTS

Numerous animals are also used in assessing the effects of environmental pollutants in toxicological research, poisons (carcinogens) that cause cancer, birth defects, and numerous other diseases. Included among toxic agents are trichloroethylene (TCE), used in spices, general anesthetics, and to decaffeinate coffee; lindane and dichlorodiphenyltrichloroethane (DDT), two pesticides; and DES (diethylstilbestrol), a chemical that promotes growth in cattle and poultry. The incredible negative influence of DDT on wildlife and the environment was the focus of Rachel Carson's famous and influential book Silent Spring.

The use of animals to study environmental poisons often produces highly questionable results. In many cases data from animals simply do not apply to humans in any direct way, and often the methods used are not adequate to draw meaningful conclusions about how pollutants affect humans. There are numerous technical and scientific problems centering on the use of animals to make predictions about human responses to drugs and environmental chemicals, many of which are related to the lack of scientific progress in helping humans deal with the negative effects of environmental poisons. Some major criticisms of the use of animals to test the effects of potentially harmful chemicals on humans include the large doses of poisons given to the animals and the fact that often the chemicals are administered in ways that do not mimic the human experience. In some studies hair dyes are fed to animals. Despite the fact that some animal tests show clear evidence of the negative effects of chemicals on the animals, the results have been ignored and the chemical industry has actually grown rather than been slowed down.

There are alternatives to the testing of environmental pollutants on animals. These include the use of human cell cultures, human living tissues, computer-based models, and human volunteer studies. Some prestigious scientists believe that the use of animals in toxicology testing is more the result of historical precedent rather than a careful assessment of the very limited utility of animal models. Philip

Abelson, a famous American scientist and former editor of the prestigious magazine *Science*, once claimed that "the standard carcinogen
tests that use rodents are an obsolescent relic of the ignorance of past
decades." These tests do not work and should be stopped.

COWS, GRAIN, AND HUMAN STARVATION

*Raising animals for consumption as human food requires a lot of food and
water for the animals who are to be eaten. A lot of land is also used to keep the
animals and to grow the grain used to feed them. For example, it takes 8 or 9
cattle a year to feed one average meat eater. Each cow needs 1 acre of green
plants, corn, or soybeans a year. Thus, it takes about 9 acres of plants a year
for the meat one person eats rather than the half acre needed if the plants
themselves were eaten. The amount of grain needed to provide meat for one
person is enough to feed about 20 people for a year. In the United States, livestock eat enough grain and soybeans to feed over a billion people. About 16
pounds of grain are needed to make a pound of beef. A reduction of meat consumption by only 10 percent would result in about 12 million more tons of
grain for human consumption. This additional grain could feed all of the
humans who starve to death each year—about 60 million people! Furthermore, about 26,500 gallons of water are needed to produce 2 pounds of beef,
whereas 2 pounds of wheat require only about 120 gallons.*

VEGETARIANISM: AN ALTERNATIVE
TO KILLING ANIMALS

*There are many alternatives for the vast majority of people who choose not to
eat meat. Vegetarian diets are much healthier than diets that contain meat,
especially meat that is been injected with various types of hormones and*

antibiotics or meat from animals who were stressed before they were killed. Furthermore, animal feces are often the source of salmonella and E. coli contamination of poultry and red meat. The Centers for Disease Control (CDC) estimates that there are about seventy-six million cases of food-borne illnesses related to the consumption of animal products each year in the United States. Nutritionist Colin Campbell, in his long-term study of dietary habits in mainland China, discovered that a low-fat (10–20 percent of total calories), plant-based diet could significantly decrease the occurrence of chronic degenerative diseases such as various cancers and heart disorders in Western countries. Yet according to the Physicians Committee for Responsible Medicine (PCRM), only one of twelve elementary schools in the United States substitutes low-fat, cholesterol-free plant protein for meat on school menus.

Due to the animal cruelty involved in meat eating, many people choose to reduce or eliminate their consumption of meat. They try cutting back on hamburgers and other animal products from five to two per week for a month, then from two to one per month, then one for two months until meat is eliminated from the diet. This practice might also help to feed other humans who might otherwise starve to death.

As Marc says, becoming a vegetarian or even cutting back on our meat consumption will help to alleviate animal suffering, help to feed hungry people, and benefit our health. Ervin Laszlow, in his new book *Macroshift,* provides some frightening statistics about meat eating. World meat consumption rose from 44 million tons in 1950 to 217 million tons in 1999. As already stated, feeding grain to cattle is wasteful—the caloric energy provided by beef is only one-seventh of the energy of the grain fed to them. Thus converting grain into beef wastes six-sevenths of the nutritional value of the planet's primary produce. On these grounds alone Laszlow suggests that "a diet based on heavy meat eating is not only unhealthy, it is immoral; it indulges a personal fancy at the expense of depleting resources essential to feed the entire human population."

Reduction in meat eating will also benefit the health of the planet. As more and more meat is demanded by the consumer societies around the world, so more and more wilderness areas are destroyed—either for grazing cattle or for growing grain to feed cattle in gigantic feed lots. The rain forests of Brazil and other countries of Central and South America have been horribly depleted as a result of the wasteful meat-eating behavior of a relatively small percentage of the world population. Yet how many people protesting the destruction of our forests stop off at a fast-food outlet for a hamburger?

Major changes in attitude do not happen fast. Whether we like it or not, we live in a world where very large numbers of people eat meat (red meat, chicken, and fish). I ate meat until I read Peter Singer's book *Animal Liberation* and learned for the first time about factory farming, veal crates, battery cages, and so forth. I stopped over twenty years ago and attribute much of my energy and good health to that decision—even though it was made on purely ethical grounds. But those who eat meat can, nevertheless, help to alleviate the suffering of animals by only buying organic, free-range products. They are not easy to find and cost more. But there is the satisfaction of knowing that the animal whose flesh is nurturing yours has had a reasonable life. (You can also feel safe from the antibiotics and hormones that contaminate meat from intensive farms.) Free-range eggs are also getting much easier to find.

Laszlow also notes that growing tobacco for export robs millions of poor people of fertile land on which they could grow cereal grains and vegetables. So long as there is a market for tobacco—and here America performs well, with a major reduction in consumption—then many farmers will plant it. And tobacco growing is very environmentally unfriendly—in Tanzania it has caused massive destruction of woodlands and forests. Reduction in meat eating and tobacco smoking will result in a better pattern of land use that will enable millions more people to be fed without destroying more wilderness and without potentially dangerous plantings of genetically modified foods. And it will greatly benefit human health.

It is easy to become depressed when we think of the terrible environmental and humanitarian problems that face us at the start of the twenty-first century—the horrifying power of the great multinationals, the wastefulness of the consumer lifestyle, the heedless destruction of the natural world, the unbelievable cruelty to animals and humans alike. Yet there is much hope. We can, by persistently voicing our concerns, upholding our beliefs, and taking action, bring about change. Once a sufficient groundswell of people truly understand the stakes and truly believe in themselves, they can achieve wonders. We can make ethical choices about what we buy and what we do not buy. And these individual choices, collectively, can change business faster than laws in our consumer society. We do not have to buy cosmetics or household goods that were tested on animals any more than we have to buy products made with child slave labor or in sweatshops. We can insist on organic food—which will be better for our health and help to keep pesticides and chemical fertilizers out of the food chain to the benefit of insects, birds, and other wild animals. And, finally, there are some inspirational role models out there, proving again and again how the action of one person can truly make a difference. That is the subject of our Ninth Trust.

THE NINTH TRUST

PRAISE AND HELP THOSE WHO WORK FOR ANIMALS AND THE NATURAL WORLD

The Ninth Trust contains stories about people who, through hard work and dedication and because of their love for animals and nature, have truly made a difference. It is important that we acknowledge their efforts. Many other dedicated individuals have helped these people along and also do important work that may go unnoticed. Let us honor these individuals as well.

Most of my adult life has been dedicated to calling attention to the importance of every individual. I believe that each one of us plays a significant role in the scheme of things, and that it is not only the individual lives of human beings that matter, but those of animals too. This is the main message that I carry around the world in my lectures, take into classrooms, and write in my books. The more I travel, the more I read, and the more people I talk to, the more I am inspired by the great potential of each human being. I am inspired by those who tackle seemingly impossible tasks and never give up, despite setbacks and criticism, until those tasks are accomplished, and by those who lead lives that motivate and inspire others.

KINDRED SPIRITS, STEWARDS, AND HEALERS

Roger Fouts, a psychologist who, along with his wife, Debbi, has extensively studied communication between chimps and humans with well-known individuals, including Washoe, recently wrote the wonderful book Next of Kin. *In his book Fouts, who deeply cares about animals, argues that by being concerned about animals and acting on issues relating to their welfare a person is more of a healer than an activist. He makes an important point—acting on behalf of animals is not only good for animals and Earth, but is also beneficial for our own spirit and makes us feel good about ourselves. And feeling that we are helping animals and the Earth, that we are healing our own and others' wounds, has motivated numerous people to take a stand and achieve many incredible accomplishments in the face of much resistance.*

In the fields of animal and environmental protection there are literally thousands of people around the globe who are devoting their lives—sometimes risking their lives—to help animals and our environment. Their efforts cover all aspects of stewardship: tireless lobbying for new anticruelty or environmental and species conservation laws or regulations; organizing or taking part in demonstrations; speaking out and establishing interest groups; and starting rescue centers for species as diverse as domestic poultry, on the one hand, and abused elephants, on the other. Rescue and rehabilitation centers are growing in number all the time as people are moved to try to take action themselves to alleviate suffering. Organizations established to enforce animal and environmental protection laws exist in almost all parts of the world.

Some projects start when one animal—a bird hurt on the road or a stray dog—is rescued, and then a second and a third, until there is no turning back. What a difference that first animal has made! Often the operation is run on a shoestring budget, and as it grows so, too, does the desperate search for funding. Always, though, there are willing volunteers to help not only in caring for the animals themselves, but also with the fund-raising and office work. And, of course, there are countless people who take abandoned, often abused, animals into their homes and become their willing slaves. There are so many moving stories of elderly people spending every spare penny on their large families of rescued cats or dogs.

In the United States it was one Henry Bergh whose energy and dedication led to the founding, in 1866, of the American Society for the Prevention of Cruelty to Animals (ASPCA)—now one of the world's largest humane societies. Bergh was particularly shocked by the treatment of horses in New York, and the first case taken to court by his new society was that of a cab driver beating his fallen horse with a spoke from one of the cart's wheels. They won the case. Bergh also worked to improve conditions for animals on farms, in circuses, and in other areas where animals were abused. As a fascinating aside, which is not generally known, he was also concerned about the cruelty to human children, and it was the animal anticruelty laws that enabled Bergh to take to court a case of extreme cruelty to a human child. He won his case, which led to the forming of the first anticruelty laws to protect children.

Henry Spira began his animal-advocacy career in 1976 by drawing attention to experiments on sexual behavior in cats that were being conducted at the American Museum of Natural History, which he succeeded in stopping in 1977. Later, in a dramatic full-page advertisement in the New York Times *on April 15, 1980, Spira opened the world's eyes to the fact that cosmetic companies routinely blind and poison animals to test ingredients in lipsticks, shaving lotions, eye shadow, and similar products. He asked, "How many rabbits does Revlon blind for beauty's sake?" Spira focused on the Draize test and the LD50 test described in connection with the Eighth Trust.*

Spurred on by the unflagging and selfless efforts of Spira and his co-workers, the anti—animal testing campaigns played a major role in forcing hospitals,

universities, and government laboratories to establish review boards to make
sure that experiments used alternatives to live animals, such as test-tube cul-
tures, where possible and to make sure that animals weren't unnecessarily
abused when they were used.

Christine Stevens founded the Animal Welfare Institute in 1951.
She has worked tirelessly on a variety of different issues relating to
conservation and cruelty to animals of all kinds. It was Christine who
perfected the art of lobbying on Capitol Hill! She and I have visited
many different legislators over the years—with a good success rate.
Another extraordinary woman is Dr. Shirley McGreal, founder and
president of the International Primate Protection League (IPPL). She
is not afraid to tackle anyone guilty of cruelty to primates or anyone
violating the international laws regarding their capture and shipment
around the world. Shirley stood up to IMUNO, an Austrian pharmaceutical giant that was keeping primates in horrendous conditions.
Like almost everyone who dared criticize it, Shirley was sued and lost
everything. But she carried on regardless and is working as tirelessly
as ever. Thousands of primates owe Shirley a huge debt of gratitude.

THE SIGNIFICANCE OF
ALL VARIETIES OF ACTIVISM

It is not always easy to get proof of the cruelty that so often takes place
behind the locked doors of the medical research establishment. Alex
Pacheco took an undercover job at a medical research facility in Silver
Spring, Maryland, to try and document alleged abuses of primates.
After two years he had his evidence in the form of secretly filmed video
footage. I remember watching this after one of my lectures—and it
made a profound impression on me, as I was new to this kind of thing
back then. It made an impression also on those whom he persuaded

to help him prosecute the laboratory. Alex went on to start People for the Ethical Treatment of Animals (PETA) with Ingrid Newkirk. PETA is now the largest animal rights group in America.

More recently, Matt Rossell took a job at the Oregon Regional Primate Research Center (ORPRC), hoping to obtain similar evidence to prove violations of federal animal welfare laws. About a thousand monkeys live in the laboratory, many of them in tiny and often filthy cages (not much more than four square feet). Matt was able to document, among other cruelties, the horror of electro-ejaculation. This technique (used during the torture of prisoners by the Gestapo) involves tightly strapping an unanesthetized adult male monkey into a restraining chair, wrapping two metal bands around the base of his penis, and applying an electrical charge. This causes ejaculation. The reason for this procedure: to acquire semen samples. The monkey who was secretly filmed (number 14609, nicknamed "Jaws" after one of his supervisors taught him to bite the bars of his cage) had already undergone the procedure on 241 separate occasions from 1991 to mid-March 2000, not counting the times when his penis was shocked more than once to get the semen sample. No wonder he tried, so desperately, to escape the ordeal. As a result of Matt's courageous undercover work, the procedure has been banned. In addition, a veterinarian has resigned and some of the scientists working there have made critical statements about conditions in the laboratory.

There are millions of monkeys and other sentient beings involved in a whole variety of procedures that, from their point of view, amount to torture. And hundreds of concerned humans trying to do something to help. Marc himself, along with other dedicated local activists, has played an important role in this arena. It was their outspoken criticism of the unnecessary use of dogs in the training of medical students that has resulted in more and more medical students opting out of physiology laboratories in which dogs are "sacrificed"—a cozy way of saying they are killed. At the University of Colorado Medical School (Marc is a professor at the Boulder campus of the University of Colorado), the number of students choosing nonanimal alternatives, which are scientifically sound and educationally perfectly acceptable, increased from two to more than forty in just four years. It takes a good deal of courage to stand up to the organization that employs you, and among his many accomplishments Marc can be very proud of that one. Fortunately, almost everywhere in the world, there are those who are prepared to risk their jobs, and sometimes their lives, to expose abuse of all kinds.

A new organization, Laboratory Primate Advocacy Group (LPAG), has recently been formed to draw attention to and protest cruel and often useless experiments on primates. This organization will be particularly effective in the future because it is mostly made up of people who used to work in the laboratories. They know only too well the kind of abuses that are so common: they can speak from firsthand experience.

An increasing number of people use their professions to try to make change. Physician Neal Barnard started the Physicians Committee for Responsible Medicine (PCRM), veterinarian Nedim Buyukmihci founded the Association of Veterinarians for Animal Rights (AVAR), and psychologist Kenneth Shapiro founded Psychologists for the Ethical Treatment of Animals (PsyETA). These three organizations have been influential in raising consciousness about the many ways in which animals are used in medical research and education that, on the one hand, cause the animals to suffer and, on the other, are unnecessary and even misleading. Physician Ray Greek and his veterinarian wife, Jean Swindle Greek, have written a book, Sacred Cows and Golden Geese, *that exposes many of the myths concerning the supposed importance of animal "models" of human disease. Kenneth Shapiro, in his book* Animal Models of Human Psychology, *has also written extensively about the lack of success of animal models in psychological research, specifically in treating eating disor-*

ders. Andrew Knight, a veterinarian in Australia, has worked tirelessly to intro-
duce humane education into the curriculum of schools of veterinary medicine.

Several philosophers have raised their voices on behalf of animals. Australian Peter Singer wrote *Animal Liberation,* describing in graphic detail abuses of animals in medical research and intensive farming and arguing for the inclusion of all sentient life in our "circle of compassion." It was this book, published in 1975, that shocked hundreds, including me, into becoming vegetarians. More recently, Peter, along with Paola Cavalieri, initiated the Great Ape Project discussed in our Second Trust. American philosopher Bernard Rollin was shocked to see so many young people who entered veterinary school as caring and compassionate gradually become increasingly callous and hard. He instigated one of the first courses in ethics in a veterinary school and then moved on to consider the way animals are regarded in medical research laboratories and the food industry. Mary Midgley has written countless books arguing the philosophical perspective for the humane treatment of animals. And two more American philosophers, Tom Regan (author of *The Case for Animal Rights*) and Dale Jamieson (with whom Marc has worked closely), have written much that has led to changes in people's attitudes regarding the moral standing of animals.

Among the most recent professions to take up the rights of animals are lawyers, a number of whom are devoting considerable amounts of time to work pro bono on a variety of cases. Gary Francione, founder of the Rutgers University Animal Rights Law Center, and Steven Wise, author of *Rattling the Cage,* are among the lawyers who have devoted large portions of their careers to the plight of animals and whose books have called attention to the numerous and complex legal issues concerning how humans treat animals. A handful of law schools also offer courses in animal law, and the number is growing each year.

Once people start to work for animals they typically continue to do so throughout their lives. Carol Johnson, confined to a wheel chair, continues her activism in an elder-care facility, where she initiated weekly discussion groups with other residents. Hope Sawyer Buyukmihci founded the 540-acre Unexpected Wildlife Refuge in southern New Jersey to care for a variety of animals, especially beavers, and continued to work until she died at eighty-eight years of age. Rita Anderson, a grandmother in Boulder, Colorado, found a new avocation in working for various campaigns to strengthen the legal standing of animals. Her grandchildren are also animal activists.

Gretchen Wyler, a former actress and committed animal activist since 1966, founded the Ark Trust to combat animal cruelty in entertainment and to promote positive coverage of animal issues in major media. It has become a major international force in animal welfare reform. Its Genesis Awards are the only major media and arts awards relating to animal issues. They are presented by the Ark Trust to honor outstanding individuals in the major media and artistic community who have communicated animal rights and animal welfare issues with courage, artistry, and integrity. In line with its major focus, the Ark Trust's slogan is "Cruelty cannot stand the spotlight."

Elly Maynard, working out of her home in Tauranga, New Zealand, was instrumental in starting what has become a worldwide movement to prevent the brutal slaughter of St. Bernards (and other dogs) for human consumption. It all began for Elly after she read about the slaughter of dogs for food. She took a petition to the streets of Tauranga and sent messages to different animal protection organizations. In the first week alone she received five hundred replies. I've had the pleasure of working with Elly and her enthusiasm is contagious, her energy boundless. "No" and "Cannot be done" do not exist in her vocabulary. Elly and he co-workers attracted the attention of world leaders, and what began as a local movement in a small city "down under" spread to about forty-five countries. Forty-five million people signed the petition that Elly helped to launch, and the International Court of Justice for Animal Rights in Geneva, Switzerland, has accepted her claim against China for importation and breed-

ing of dogs for food. Dogs are better off because of Elly's and others' efforts. Elly is now president of the Phoenix Animal Charitable Trust, Inc. There always is something to do for animals.

ONE DOG AT A TIME

We cannot always tell how our actions or our words may affect others. A few years ago, during one of my interminable lecture tours, I spoke in Kalamazoo, Michigan. During the talk I described an incident that had just taken place in Kigoma, Tanzania. I'd seen a group of small boys teasing a puppy. I went to remonstrate with them—at which a European friend told me I would be wasting my time. "That kind of thing goes on all over Tanzania," he said. But, I thought to myself, *this* is happening right *here*. I asked the boys if they would like to be treated as they were treating the puppy? No. Did they think the puppy liked it? No. So why were they doing it? They did not know. "Your puppy is in good condition. You like her, don't you?" I said. They admitted they did! So I told them about our Roots & Shoots program, and asked if they would like to help. They went off, very excited, carrying the puppy.

Two years later, at a conference in Shanghai, I was approached by one of the conference participants, excited to meet me. She told me how she had been posted to a school in Bombay and, after the first couple of terms, was just about ready to resign. "I couldn't stand the plight of the street dogs," she said. "They were everywhere, and they were mangy, sick, and starving." By chance, her home was in Kalamazoo, and when she went on leave she confided in an old friend of hers, a veterinarian. He had attended my lecture and told her the story of the boys and the puppy. "Why don't you go back," he told the teacher, "and take it one dog at a time." And that is what she did. She rescued one, bathed him, gave him his shots, and neutered him. Then she found him a home. Then she did it all over again. And again. One of her friends, inspired, decided to do the same. Then another friend joined their work. Between them they started a shelter. Gradually, imperceptibly at first, they began to see a difference. The strays were disappearing from the streets. One dog at a time.

I visited an amazing dog and cat shelter in Beijing. It was founded by a remarkable Chinese woman, Zhang Luping, who runs it with the help of many dedicated volunteers—they take turns sleeping there so the animals will not be lonely. There is a "no kill" policy. And there are no "kennels" in the traditional sense, but open spaces with trees and grass where the dogs and cats are free to play together (in their separate enclosures). It was a visit arranged by JGI–China, and I was with a group of schoolchildren who were learning how to groom and love the dogs. What a wonderful sight to see dogs and children loving each another—in a country in which people are also known to eat dogs. A similar shelter, with trees and open space, was started for dogs by a group of students on their university campus in Taiwan.

Marc and I get so many pleas for help from people struggling to make a difference in their communities. Renee Andrako called me, sounding desperate, from Idaho. She was doing all she could to help the plight of the dogs in her neighborhood, many of whom are horribly abused. The laws regarding cruelty to animals are weak and seldom enforced. She and her dedicated co-workers not only care for the rescued dogs, but have to work full-time to support their efforts. I gave Renee Marc's e-mail address, and he offered her some help and encouragement. We only wish we could do more.

ANIMAL SANCTUARIES

We get much inspiration from those who devote their lives to the betterment of life for animals. Many sanctuaries in different parts of the world care for farm animals who need help. In 1996 Deanna Krantz founded the India Project for Animals and Nature (IPAN) in South India, which seeks to protect local animals, including donkeys, dogs, cats, ponies, water buffalo, cattle, sheep, goats, rabbits, and orphaned wildlife. This project operates Hill View Farm Animal Refuge, a home for more than two hundred abandoned, abused, and injured animals. Michael W. Fox, a bioethicist with the Humane Society of the United States, is IPAN's veterinarian and chief consultant. Krantz and Fox have both faced and overcome innumerable frustrations and dangers. Another wonderful program in India is the Wildlife Trust, founded in 1998, where working elephants are given free care and local veterinarians and mahouts are trained in elephant health and management. Carol Buckley and her staff provide homes for elephants at the Elephant Sanctuary in Hohenwald, Tennessee. Their goal is to give back to elephants what was taken from them when they were taken from their natural environs and their families.

Gloria Grow and Richard Allen established a refuge for farm animals around their house near Montreal, Canada. Later they became obsessed by the plight of chimpanzees in medical research. They built a sanctuary for fifteen of the three hundred chimpanzees retired from the Laboratory for Experimental Medicine and Surgery in Primates (LEMSIP) colony. It was a huge task, but, successfully completed, it gave courage to others who wanted to help. A few years later Carole Noon founded and runs the Center for Captive Chimpanzee Care in Florida. This sanctuary houses twenty-one of the so-called air force chimps, part of the colony that was used for research by the United States Air Force in its space research program. The air force has been leasing these chimpanzees to various research laboratories, including the Coulston Foundation in New Mexico—a facility that has been convicted of more cases of animal abuse than any other. Another sanctuary was built by Patti Regan for a growing number of chimpanzees and orangutans discarded by the pet and entertainment industries.

In the UK, Jim Cronin founded the Monkey World Ape Rescue Centre, which he now runs with his wife, Alison. Originally this center was built to provide a permanent home for the infant chimpanzees who were smuggled into Spain from West Africa and used as photographers' props in tourist resorts. Jim worked with a British couple who lived in Spain, the Templar's, and with the police, to stop the illegal trafficking, and also with tourist agencies, persuading them to warn visitors of this despicable practice—the photographers used lion cubs and other animals as well, all of which were treated with great cruelty. Monkey World has rescued chimpanzees and orangutans from many parts of the world. And then there are the tireless conservation efforts of Biruté Galdikas, who has worked for decades to stop the decimation of orangutan populations due to habitat loss as a result of logging, mining, and the illegal animal trade. Likewise, Willie Smitts, whose group at the Wanariset Forestry Station in Indonesia cares for orphaned orangutans and helps them learn to survive in the wild so that they can be released.

In Africa there are literally hundreds of orphaned apes—chimpanzees, gorillas, bonobos—and many other animal species as well. The desperate situation is mostly the result of the commercial trade in bushmeat. Mothers are shot for food, but there is little flesh on an infant, who is offered for sale in the market. Because this is illegal, wildlife authorities can be persuaded to confiscate the youngsters—but they must be cared for. Usually they arrive in pitiful condition, malnourished, hurting, and traumatized. Even when they have been nursed back to health, they cannot be returned to the wild—they will probably have to be cared for for the rest of their long lives. A handful of extraordinarily dedicated people work in the growing number of ape sanctuaries in various African countries. The first of these was established by Dave and Sheila Siddle, with their own funding, on their cattle ranch in Zambia. It all started when one tiny and badly wounded infant was confiscated from a hunter who had brought her over the border from what was then Zaire. As of 2001, the Siddles were caring for eighty-five chimpanzees.

One of the major problems of working in Africa, far more serious than the lack of infrastructure, the expense, and the corruption, is the danger. JGI's largest sanctuary, Tchimpounga, is in Congo-Brazzaville (The People's Republic of Congo), a country that has suffered nine years of civil war. Our project manager, Graziella, has been in life-threatening situations on a number of occasions, including confrontations with groups of armed militia. She has been yelled at, threatened, and once physically thrown out of her house. But she is still there, dedicated to her enormous task of caring for a hundred orphan chimpanzees.

Bela Amarasekaran and his wife risked their lives over and over again during the terrible civil war in Sierra Leone. They stayed in Freetown and somehow cared for the infant chimpanzees who had already arrived in their new sanctuary before the start of the war. Betsy Brodman, working for the chimpanzee colony belonging to the New York Blood Center, stayed on, with her husband, in war-torn Liberia, creeping out to find food for the chimpanzees despite the danger. After watching her husband shot and killed as he knelt in front of a rebel soldier, begging for his life, she was persuaded to leave. But not for long—she was soon back and again caring for the chimpanzees.

Alex Peal also risked his life during the Liberian civil war. Alex was Director of National Parks when the fighting began. He got his family safely out of the country, but returned to try to continue his work. He was captured by rebel troops but, incredibly, as a result of desperate appeals from the international conservation community—Charles Taylor intervened personally—Alex was released. Eventually he got to the United States, where he worked on a variety of conservation issues until the day came when he was able to return to Liberia.

Another individual who has faced danger in Africa on many occasions is Swiss photographer Karl Ammann. It is thanks to his efforts that world attention has been drawn to the bushmeat trade, the most urgent conservation crisis of our time. And he did this by traveling into the most dangerous places, collecting photographs that would compromise logging company employees, government officials, and hunters—all engaged in illegal activities. Fortunately, he suffered no more than loss of a camera and a few nights in jail. Michael Fay has also risked his life in his work to conserve the Eden-like Endoke forest. Recently he walked 2000 miles across Central Africa to find out, for himself, the state of the forests and their wildlife. Dr. Richard Leakey, son of my mentor Louis, has also run a dangerous course in his efforts to protect wildlife in his beloved Kenya. With extraordinary courage he challenged, as director of the Kenya Wildlife Service, corrupt officials in many government departments and exposed illegal activities on the part of some of the most senior of them. It is due to Richard that the elephant population in Kenya rebounded after being all but exterminated by poachers.

It is not only in Africa that people have continued to help animals in the face of civil unrest and war. One example comes from Afghanistan. In the Kabul Zoo, the director, Sheraga Omar, and a handful of keepers somehow managed to keep some of the animals alive throughout the grim Taliban years. When asked why he did this, Omar simply said: "We cannot let these animals die." The most famous of the zoo's long-time residents was Marjan, a lion who lived there for twenty-three years before dying in January 2002. Marjan survived a grenade attack despite losing one eye and part of his jaw.

Other animals also have been targets in Kabul's zoo and the situation is dangerous for humans too. A twenty-five-year-old female elephant was struck by a rocket-propelled grenade and killed. About the horrible conditions, Omar says, "We have hope . . . we are running on faith."

TAKING ON LARGE ORGANIZATIONS

In these days of intensive, or factory, farming, a growing number of people are trying to help the cows, pigs, poultry, and other food animals who suffer so greatly. During the 1980s, Henry Spira turned his attention to these issues. He focused on working with the meat producers, trying to persuade them to reduce animal suffering. He also worked with some of the multinationals—such as McDonald's—urging them to insist on acceptable standards of husbandry in the operations that supplied their meat. He made some headway—but we have a long way to go. Pioneers in designing more humane conditions for pigs were Professor David Woodruff, of Edinburgh University, and his student, Alex Stolba, who continued the work at the University of Zurich. In Australia Patty Mark founded and runs Animal Liberation Victoria and is working to outlaw the use of poultry battery cages. And we could go on and on listing the people who not only care, but work to make a difference.

Perhaps the most daunting task for those who want to change the status quo is the thought of confronting some of the giants of industry. In our Sixth Trust we described the determination of Rachel Carson as she tackled the petrochemical corporations. In an extraordinary act of defiance, two Britishers, Helen Steel, a gardener, and Dave Morris, a postman, decided to sue McDonald's. The McLibel case, as it came to be called, was one of the longest trials ever. On June 19, 1997, it was ruled that Helen and Dave had not proved the allegations against McDonald's concerning rain forest destruction, heart disease and cancer, food poisoning, starvation in the Third World, and bad working conditions. However, they had proved that McDonald's "exploits children" with its advertising, that it falsely advertises foods as nutritious, that it risks the health

of its most regular, long-term customers, and that it is "culpably responsible"
for cruelty to animals. McDonald's conceded in writing on January 5, 1999,
that the trial judge had been "correct in his conclusions."

Another important victory was won against the USDA by J. J. Haapala, an organic seed grower and director of education for Oregon Tilth. It all began when the USDA attempted to change the conditions for growing organic food. It was proposed that genetically engineered products, some pesticides, and sewage-sludge fertilizer could all be used on organic farms. I well remember my own horror when I learned about this—indeed, I wrote a personal letter to the Secretary of Agriculture. The campaign was successful, and now any farm that wants to be certified as organic must comply with new regulations that ban, among other things, the use of biotechnology and radiation.

Some individuals have fought long, hard, and successfully on a variety of conservation issues. One such was the late Margaret Owings. In 1970 I stayed at her house, which was built into the cliffs above the Pacific Ocean, close to Big Sur. From her sitting room window I looked out at the rugged, unspoiled shoreline—and I knew that it was entirely due to the lobbying and fund-raising efforts of Margaret and her late husband, Nat, that it was not another Malibu. They started this amazing conservation program long before the mainstream environmental movement got going. It was Margaret's voice and persistence that prevented old-growth redwood forests from being paved over. She fought to prevent the meandering Pacific Coast Highway—along which I had driven that day—from being blasted and straightened into a four-lane freeway.

But she is perhaps best known for her efforts on behalf of the California sea otters. As we sat together in her house, she showed me where some sea otters swam. She had been campaigning to protect them from the vindictiveness of the abalone fishermen—who blamed the otter, rather than their own overharvesting, for the depleted stocks. Margaret was an artist and writer, not a biologist. But in 1966 she created Friends of the Sea Otter and, with a small band of dedicated volunteers and the support of the media, she succeeded in building up a public image of these utterly charming animals, who bob about on the sea, lying on their backs among the kelp forests and using rock tools (which they keep under their arms when diving) to hammer open abalone and other seafood that they lay on their bellies. Suddenly sea otters were the darlings of the public, who drove long distances to see them. Thanks to her efforts, legislation was introduced making it illegal to shoot them, and the southern sea otter made a gradual comeback from near extinction.

As early as 1962 Margaret had turned her attention to the protection of the cougar, or mountain lion. It started when she heard a shot, went to investigate, and found a young man posing for a photo, his foot on the dead animal. Eventually, in 1989, her campaign resulted in Californians' doing something unprecedented: they voted to ban cougar hunting permanently, taxing themselves to protect habitats. California is still the only state to protect these beautiful cats.

Margaret Owings exemplified the "power of one." It has been said of her by former state Senator Fred Farr, "Without Margaret, California would have awakened a lot later." Just before she died, she was asked how she would like to be remembered. After much thought, she replied, "She cared."

Margaret Owings, with her love of the redwoods, would have enjoyed meeting Julia Butterfly Hill, who, on December 10, 1997, climbed into a 180-foot-tall California coast redwood tree she named Luna. Her aim was to prevent the destruction by logging of the tree and of the forest in which Luna had lived for a millennium. Julia came down from Luna on December 19, 1999. During the two years she spent in Luna, Julia attracted worldwide attention for her non-violent action in defense of the forest. She was an inspiration for thousands of people. Her actions resulted in her becoming the youngest person inducted into

*the Ecology Hall of Fame. Julia went on to establish the Circle of Life Founda-
tion, which promotes efforts to protect and restore Planet Earth.*

Mr. Lian Congjie is a retired professor of history in Beijing with a
deep love of nature. He started an organization, Friends of Nature,
and was instrumental in raising the awareness of the general public
long before the recent environmental disasters—the dust storms and
flooding—brought such issues into the mainstream Chinese media.
Friends of Nature produces written information and videos about a
wide range of natural history topics—and it served as the foundation
for the modern environmental movement in China. It was Lian who
was responsible for bringing the plight of the Tibetan antelope to
world attention. The last time I saw him, one of his best friends had
just been murdered by poachers for his brave efforts to halt the
slaughter. Organizations like Friends of Nature are few and far
between in China, and only someone with enormous dedication and
perseverance could have created and sustained it. It has taken
courage, too, as some of the issues he has addressed were not always
popular with the government.

John Hare is very English. During his time in foreign service he
developed a taste for adventure, so was quick to sign up, when he had
the opportunity, to go on an expedition to the Lop Nur desert in
China, one of the harshest deserts in the world. He survived all manner of hazards, and he also became utterly absorbed in the plight of
the last remaining wild Bactrian camels—those with two humps. Only
some 260 of them survive in the wild in China, with another slightly
larger population in the contiguous deserts of Mongolia. These
camels are genetically quite distinct from domestic Bactrian camels.
With extraordinary persistence John managed to get together funding for a series of expeditions. On one of these, he and his small team
navigated the hostile desert on domestic Bactrians to survey the
migration route of the wild camels. Amazingly, he was able to per-

suade both the central and local governments to agree, in writing, to the creation of the Lop Nur Camel Sanctuary, one of the largest national parks anywhere. And, even more extraordinary, he has just overseen the signing of the first joint agreement on any conservation issue between the Chinese and Mongolian governments. Now the camels will be protected on both sides of the border between the two countries.

OTHER INSPIRATIONAL ACTIVISTS

It is not necessary to start a movement, create a national park, or influence a government in order to make a difference. When my mother was in her late seventies, she went into one of the big super-markets in Bournemouth, UK. When she went to get eggs, she looked for a sign indicating free-range eggs. She couldn't find one, so she went over to a young assistant and asked if they had any. "What's a free-range egg?" the girl wanted to know. Vanne explained that it was laid by a hen who could move about and scratch about on the ground. "Don't all hens do that?" asked the assistant, puzzled. So Vanne explained about battery farms, about debeaking and how blood oozed from the mutilated beaks. How claws—and sometimes the end of the toe bones—were trimmed so that the claws did not grow and become entangled in the wire. The young woman looked horrified and a small crowd of customers gathered round to listen. The assistant manager appeared to see what was going on and led Vanne into an office at the back of the store. There she repeated her story. The following week—and ever after—free-range eggs were on sale. And then she did the same at Sainsbury's.

Jon Stocking, as a young man wanting to see the world, got a job on a tuna-fishing boat. It was before most people were aware that dolphins lost their lives in the fishing nets. One evening, as the nets were being pulled in, Jon looked down over the rail and saw an infant dolphin being crushed against the side of the boat. Close beside it was an adult—the mother he assumed. She seemed to be looking at him with an appeal for help. He found himself jumping into the water—among a mass of huge, terrified tuna, dolphin, and sharks. Somehow, Jon managed to lift the infant over the net to freedom. Then he had to struggle to save the mother. As he watched mother and child swimming into the sunset he suddenly realized that seven or eight dolphins were lined up. So he cut the net with his pocketknife, thus freeing the tuna along with the dolphins. Jon lost his job!

The incident haunted him. He realized how many other beautiful creatures were in trouble. At last he found a way in which he, personally, could do something about it. He got another job on a boat to Belgium, where he could learn to make the best chocolate. He started his own company, the Endangered Species Chocolate Company. He donates at least 10 percent of the profit, before the deduction of tax, to an organization working to save the animal depicted on the chocolate wrapper.

There are thousands of relatively small programs around the world, started by individuals who care, individuals with the determination that can lead to success. In certain parts of England, during the spawning seasons of toads, hundreds are killed as they cross busy roads to get to their traditional breeding grounds. Someone had the idea of creating a series of tunnels under the roads. Every year, in the breeding season, dozens of volunteers report for duty, guiding the toads to the tunnels—and then ensuring that they return safely when the eggs are laid. There are tunnels for the native ponies and deer under the main roads through the New Forest. They were dug when fences were erected to prevent ponies from wandering onto the roads and being killed—fences that prevented them from traveling between their traditional grazing grounds. This happened only when a small group raised awareness, wrote letters, and raised money.

In Kenya a biologist found that the red colobus monkeys she was studying were getting killed when they came down from the trees in order to cross a new road. So she raised the money to construct rope bridges crossing high overhead. These were immediately successful and other species of monkeys used them too. In Banff National Park in Canada, it was found that many animals used the tunnels built under the main access road for tourists. But the Big Horn sheep were terrified to go underground, even for such a short distance. When dizzyingly high and perilous-looking bridges were constructed *over* the road, the sheep crossed with complete confidence. Recently, rope bridges have been strung over the roads in parts of England to protect the endangered dormouse. If more people lobbied to make tunnels and bridges, it would make a huge difference for all the wild animals whose ancient pathways are bisected by roads and ever more roads. In some areas the number of roadkills only decreases when most of the animals living there have been killed. Only some species develop good road sense.

Making a tunnel or bridge is a kind of compromise. You cannot get rid of the road, so you find another solution, less good but better than nothing. One of the most imaginative compromises was the brainchild of Ms. Ing, managing director of the Rapid Transit Company in Taipei, Taiwan. Plans were already well developed for the building of a railroad to link the north and south of the island when it was discovered that the planned route would utterly destroy the larger of the two remaining breeding grounds of the Taiwan pheasant-tailed jacana—considered by many Taiwanese to be their national bird. They breed in wetlands—which were once widespread. But today many of these areas have been drained in Taiwan, as elsewhere, and there has been serious decline in the numbers of jacana over the past fifteen years or so. Local ornithologists, distressed at the disastrous effect of the new rapid transit development, contacted concerned scientists from around the world. The company went back to the drawing board to search for an alternate route. They failed to find one. And so Ms. Ing decided that if they could not move the railroad, they would move the breeding ground! The company bought up large areas of land that had been drained and reclaimed. The water was routed back and the wetlands soon returned. The great expanse of shallow water was planted with water chestnuts. And everyone held their breath.

I visited the site just after the first breeding season. There was jubilation all around—more pairs had bred in the new refuge than in the old breeding ground, and more chicks had hatched. A visit by the new president was arranged to coincide with my visit to Taiwan. And, through binoculars, after speaking to the press, we looked at the elegant long-legged birds as they fed in their new pastures—a perfect example of a win-win situation. And the irony of it is that, but for the plan to build the railroad through their habitat, it would have gradually become smaller and smaller as more and more land was drained. That is what had been going on for years. And probably no one would have noticed until it was too late!

MOVING FORWARD

We have described the work of some very inspirational and dedicated people—individuals whose efforts are making this a better world for all living beings. Many have focused on local issues, but have had much wider impacts. The more success stories we hear, the more we are ourselves motivated to help. Each and every lone voice matters. It is important to act.

In the post–September 11 world, animal and environmental causes need our help more than ever. They are no less important just because human need is so great. That is why it is so meaningful when celebrities from theater and cinema and television take on some animal or environmental cause. Pop stars such as Bono, Sir Paul McCartney, and Sting, television producers such as David Attenborough, and film stars such as Alec Baldwin, Kim Basinger, Candice Bergen, Pierce Brosnan, John Cleese, James Cromwell, Harrison Ford, Whoopi Goldberg, Angelina Jolie, Robert Redford, and Alicia Silverstone lend their names to fund-raising efforts as well as speak out for the causes they believe in. They are raising consciousness, they are making a difference.

We must remember one thing—however impressive the work of an inspirational leader, he or she will not accomplish much unless an army of dedicated people join the effort. Together we can make a difference. Together we can bring about change. That is the theme of our last Trust.

THE TENTH TRUST

ACT KNOWING WE ARE NOT ALONE AND LIVE WITH HOPE

The Tenth, and final, Trust is, perhaps, the most important of all. It reminds us that every action we take to make the world a better place is important and worthwhile, no matter how small. Because there are millions of others like us, and as long as each of us does our bit, the cumulative result will be massive change for the good.

Although we often pay lip service to the importance of individual action, it is sometimes very difficult to believe that our own behavior can really make a difference. We want to see change in the world, to try to live in harmony with nature; we know how we could be helping to make the world a better place for animals and humans alike. But we feel small and insignificant in a world of more than six billion people. How can the actions of one person *really* make a difference in the often selfish, uncaring, unfeeling world around us? We are, after all, but tiny cogs in a giant machine—a machine that is relentlessly destroying that which we love, bit by bit. And so we sink into apathy. And so we do nothing. But there is hope. We still have a chance to turn things around if only enough of us understand the importance of our own actions. And we must *act* . . . and act *now*. With new hope, knowing that we are not alone in caring, that we are part of a growing band of people determined that, together, we can succeed. How do we shake off this apathy? Hope comes from knowing that we are not alone in our caring, that we are one of a growing band of people determined that, together, we shall truly make the world a better place.

GLOBALIZATION

Some aspects of globalization chill my soul. The ruthless exploitation of the poor—the "theft" by the wealthy of profit reaped by the sweat of peasant brows; the spreading of materialistic Western lifestyles; the insistence that only wealth and a high standard of living are important; and the gradual erosion of cultural diversity. And the hypocrisy. All of these chill my soul. But globalization is not all bad. There are companies that respect the cultures where they work, that give fair compensation to their workers and improve their lives by building hospitals and schools. And, best of all, there are the new channels for communication that are gradually linking citizens around the globe. Suddenly people struggling in tiny groups to try to improve their world can get in touch with people in other places who are trying to do the same thing. Suddenly they realize that, together, they are making a big impact. And once a project is perceived to be successful, others join in.

Small groups of people who are trying to improve the lives of animals, to introduce animal protection laws, to strengthen antipoaching regulations, and so on, are springing up in eastern Europe and in developing countries. And now, for the first time, they have the possibility of communicating with others who feel as they do in different parts of the world. And they can share information and experiences, get support and sometimes financial help. Suddenly things do not look so bleak. Indeed, the new information channels open up whole new possibilities for collaborating for change.

Marc and I, like so many others, are using the Internet to help our cause. Long ago, when I was still working for my Ph.D. at Cambridge University, I was shocked by the callous treatment of living animals by ethologists interested in how things worked. The more I read about the kinds of research going on in laboratories around the world, the less I wanted to be a scientist. In the name of science, birds were deafened, monkeys were strapped into restraining chairs—sometimes for days— with electrodes in their brains, kittens had their eyelids sewn shut, baby monkeys were taken from their mothers and raised in isolation, dogs received electric shocks, birds were starved to death to see how long they could live without food, rats were forced to swim until they all but drowned to see how long they could survive—the list is endless. Over the years a number of organizations have been formed by scientists trying to protect animals exploited in their particular discipline, but ethologists were not represented. And I could not find anyone who was interested in starting such an organization until I met Marc.

ETHOLOGISTS FOR
THE ETHICAL TREATMENT OF ANIMALS

In June 2000 Jane and I formed the international organization Ethologists for the Ethical Treatment of Animals/Citizens for Responsible Animal Behavior Studies (EETA/CRABS). Originally the group was simply called Ethologists for the Ethical Treatment of Animals, but there was so much interest from teachers, lawyers, students, veterinarians, wildlife rehabilitators, behavioral consultants and others, that we added "Citizens for Responsible Animal Behavior Studies."

Our purpose is to develop and maintain the highest ethical standards in comparative ethological research conducted in the field and in the laboratory. People are urged to use the latest developments from research in cognitive ethology and on animal sentience to inform discussions and debates about the practical implications of available data and for the ongoing development of policy. Bringing together and working with people from various fields, with different backgrounds and interests, but with shared visions, raises awareness and fosters communication, so that ethological research will come to be conducted more ethically and responsibly and alternatives to routinely invasive methods will be developed and implemented.

Already one graduate student has changed her Ph.D. research to be more humane, and we have called attention to and had people reprimanded because of various forms of animal abuse worldwide. We have also called into question projects in which chimpanzees are cross-fostered to be raised as humans, for there seems little to be gained by repeating research that was performed previously. Also, because of our organization, people will be made aware as quickly as possible of approved alternatives. We recognize that ethological research will continue in the future, and we believe that we must be held accountable for how we study animals. As we learn more about the cognitive and emotional lives of other beings and this information is shared widely, windows into their lives will be opened and a deeper understanding of their minds and emotions will help us develop more ethically sound, noninvasive methods.

Thousands of people have visited the EETA/CRABS Web site. Suddenly a student writing with an ethical problem realizes she is not alone. There are others who feel the same—in numerous countries. There is a professor who has felt the same and has worked out a solution. The problems—and the solutions—are shared. Each voice adds a new note of determination; helplessness turns to hope.

MAKING A DIFFERENCE: THE IMPORTANCE OF CHILDREN

As I travel around the world I continually meet young people, especially college students, who are very bitter about the future—they feel that we have compromised it. And they are right. When I think of the world as it was when I was born in 1934 and the world into which my grandchildren have been born, I am filled with shame. So much has been damaged, some of it irrevocably. So many species have become extinct. Which means, of course, that they are gone forever.

Some of these young people have become angry and this has sometimes led to violence. Some have developed "do not care" attitudes—let us "eat, drink, and be merry, for tomorrow we die"; the sort of attitude that led to the wild partying that went on in England during World War II. Other young people have become withdrawn and apathetic. Some are seriously depressed. It was my aching concern for youth that led to the birth, in 1991, of the Roots & Shoots program. Roots & Shoots, as mentioned in connection with earlier Trusts, is a name that symbolizes the often hidden power of young people: the power to heal the planet, to gently but inexorably break through all obstacles toward a brighter future.

Wherever there is a Roots & Shoots group, the members learn more and more about animals and the natural world. One of the motivating factors underlying the creation of the program was my dismay when I found that Tanzanian secondary-school students were learning nothing about the wild animals that make their country so attractive to tourists from other countries—yet they were hungry for such knowledge. Everywhere it is the same—children want to learn about animals and nature—all around the globe. Indeed, it is universal. The tragedy is the number of children in our ever growing urban areas who are no longer connected to nature, who are condemned to a childhood of concrete and steel. And it is just as bad in places where deforestation and poverty blight the life of children and dull their minds. Still, once we can help young people to understand the problems around them and help them to start taking simple actions that will lead to a change for the better, we begin to see a difference in their attitude. Roots & Shoots is about building self-esteem, empowering youth to work together to make their world a better place.

It is the energy and enthusiasm of youth, the love for life that lies in each heart, even though it may lie dormant for a while, that spells hope for the future, including the future of some of our most endangered species. Two Roots & Shoots groups in Beijing are learning about the river dolphin that is on the verge of extinction—only twenty individuals remain. The children are raising money to contribute to the desperate rescue project. There are only about twenty Channel Island foxes in California—they too have a Roots & Shoots group raising money, spreading awareness, fighting to save them.

Another group is trying to help save the snow leopards of the Himalayas, raising money for building fences that will keep sheep safe from hunting leopards at night so that the farmers will not need to shoot them. And everywhere groups are restoring habitats, cleaning streams, building nest boxes for all manner of birds, including owls. And for bats and even solitary bees. Setting out feeders, and string bags full of nesting materials. And watching, and learning.

A group in Germany is working to protect field mice living on farmland. The mice are being poisoned by farmers because they damage the crops. The Roots & Shoots group is observing these mice and finds that they do much less damage than stated. They are protesting the poisoning because it is affecting the birds of prey and other natural predators. A group in Chicago heard about plans to build a plant for bottling water from a local spring. They researched the effects of such a development on the ecology and wildlife downstream, balanced it with the creation of new jobs, and presented such a strong case against Perrier that plans were halted, pending an official environment assessment study.

Groups work to create or maintain nature trails. A group of high-school students is spreading awareness in the community about the dangers of water pollution to wildlife. Another group is explaining the danger to wildlife caused by helium-filled balloons—large fish and some marine mammals swallow the deflated balloons that end up in the ocean and die agonizing deaths. Other groups are fighting to get rid of the plastic holders that bind together our six-packs of beer and soft drinks. Some are spreading awareness about the need to cut the plastic, explaining how many creatures get them caught around their necks and get strangled. A group in Japan is volunteering much needed help in one of the very few refuges for animals in that country.

Elsewhere, children help in rehabilitation centers. Many groups are volunteering to care for animals in shelters and helping to find homes for unwanted strays. A Tanzanian group is investigating conditions for goats and chickens in the markets. Another in America is working with abused horses; and one in the UK rescued hens, utterly without feathers, from a battery farm after lives of overproduction and restriction of movement and studied their gradual rehabilitation. Thousands of children are learning how to care for dogs and cats, guinea pigs and rabbits, chickens and cows. Groups in America, Europe, China, Singapore, Thailand, South Korea, Mexico, and the Congo, in collaboration with keepers in their local zoos, design and implement ideas to alleviate the terrible boredom of life in a cage. Learning about the animals in the wild, making observations in the zoos, and discussing how they think the animals feel is an important lesson.

Hundreds of groups are organizing fund-raising events on behalf of a wide range of animal protection issues. Hundreds of animals have been "adopted"—including many of the chimpanzees in our JGI sanctuaries. Marc's niece, Nicole Morse, who lives in Weybridge, Vermont, makes clips out of beaded dolls attached to colored barrettes and sells them at local stores. The profits go to JGI for chimpanzee guardianships

Several members have asked their friends to donate money for the JGI-chimp projects in lieu of birthday presents. India began saving money for our chimpanzees when she was five years old. Four of her friends helped. By earning money, then asking parents and other adults to match their earnings, they were able to present me with checks worth more than $2,000 when India was seven years old. (She had kept some money back so that she could use it to make more!) Indeed, one individual makes a difference.

One group demonstrated peacefully against a particularly cruel (and unnecessary) series of medical experiments involving rhesus monkeys. Members of a Roots & Shoots group in Boulder, Colorado, are working to get the state legislature to pass a bill that would require public schools to provide nonanimal alternatives to dissection. Members of other Roots & Shoots groups protested against the planned poisoning of prairie dogs on the campuses of five schools in Boulder County. They also volunteered to help move these wonderful, family-oriented rodents to save them from being killed. And there are hundreds of groups that are writing letters, lobbying for or against an infinite variety of causes.

So many of the young people who take an active role in Roots & Shoots are quite extraordinary. Caitlin Alegre spent a whole weekend, at the age of nine years, writing out notices telling people not to buy Proctor & Gamble products because of its animal testing. Caitlin went on to initiate a Roots & Shoots group at the Northern Lights School in Oakland, California, which, among other projects, researched a whole variety of products and advised parents and friends about those they should and should not purchase. Caitlin and some of her friends subsequently formed a band called the Envirochicks. They compose music and write lyrics about animals and the environment. They are very professional now and deliver inspirational messages about how we can help Planet Earth.

Lisa Thomas is another remarkable individual. After one of my talks in Johannesburg, South Africa, she determined to start a Roots

& Shoots group at her school. But she met a wall of resistance from the staff and a disappointing lack of enthusiasm among her peers. And so, while still working her way around these hurdles, she decided to start doing something herself. Near her home is a big shelter for stray dogs, a huge problem in South Africa. Once a week many of those who have not been claimed are euthanized. Lisa, only fifteen years old, was haunted by the thought of all those dogs, some coming from homes where they had been loved, shut into small, smelly cages, abandoned and heartbroken. She decided she would try to bring love to them at the end of their lives. And so, once a week, Lisa goes to the shelter and spends at least ten minutes with every dog who is going on his or her last journey. The moral strength of that young woman in truly incredible—I do not think I could do it, week after week.

Roots & Shoots has a gentle nonviolent philosophy eschewing violence, believing that the tools for change are knowledge and understanding, hard work and persistence, love and compassion leading to respect for all life. It is desperately important, in this era of rapid globalization and more especially after the shocking events of September 11, that children learn to understand and respect individuals from different cultures, ethnic groups, religions, socioeconomic levels, and nations.

An important component of the Roots & Shoots program is Partnerships in Understanding, which links groups around the globe. It helps them to realize that they are not alone, no matter how remote the place where they are located. Their partner groups feel the way they do, care as they care, and are as determined as they are to change the world. It is desperately important to ensure that young people have hope for the future, that they develop self-confidence, self-respect, and respect for others. For other human beings. For all living beings. We believe passionately, Marc and I, that an understanding of and respect for animals can be a key ingredient in the makeup of a loving and nurturing human being.

All of these activities are changing attitudes—not only of the children, but often of their families and teachers as well. Add to this the thousands and thousands of children around the globe who are working on similar projects outside the Roots & Shoots program, and we feel even more hopeful for the future. And the Internet links them together, allows them to express themselves, provides a platform and an audience for their hopes and fears, a way of sharing their excitement and disappointment. The children are learning compassion and respect and how to work gently but firmly against obstacles. Learning to never give up.

THE IMPORTANCE OF REMAINING POSITIVE AND HOPEFUL: THE POWER OF EYES

People often ask Marc and me how we manage to keep a positive outlook despite the terrible things that are happening in the world. We always tell them that we get inspired and encouraged when we meet our Roots & Shoots groups, full of pride in their achievements, their eyes shining, knowing that they have made a difference in their own community. And knowing that they are part of a worldwide family of caring, compassionate, and dedicated young people. The citizens of tomorrow's world, prepared to work hard to preserve what is left, restore where they can, and live in harmony with nature everywhere.

One story I have told before—very many times. Yet I must tell it again, for it symbolizes all that Marc and I are trying to say when we talk about the relationship between human and animal beings. It is about a chimpanzee called JoJo, who was born in Africa. When he was about two years old, his mother was shot and he was taken from her bleeding body and shipped to America. For many years he lived alone in a small, barren cage. Eventually money was raised to build a large enclosure, surrounded by a moat (since chimps cannot swim). Nineteen other chimpanzees were purchased, introduced to each other, then released into the enclosure.

One day one of the other males challenged JoJo, who ran into the water. He managed to scramble over the fence intended to stop the chimps from drowning in the deep water beyond. Three times JoJo surfaced, gasping for air, and then he was gone. On the other side of the moat was a small group of people. A keeper ran to get a long pole. Luckily for JoJo, a zoo visitor named Rick Swope was there with his family. He takes them one day each year. Rick jumped in and swam, feeling under the water, until he touched JoJo's inert body. Heaving the dead weight over his shoulder, he scrambled over the fence, pushed JoJo onto the shore of the exhibit, and started back toward his family.

Suddenly the human onlookers began screaming at Rick to hurry. From their elevated position they could see three big males, hair bristling, moving toward the scene. At the same time JoJo was sliding back into the water—the bank was too steep. A woman happened to capture the scene on video. We see Rick standing by the fence. He looks up toward his family, then to where three males are approaching, then down at JoJo who is just vanishing into the water again. For a moment, Rick is motionless. Then he goes back, again pushes JoJo up onto the land, and waits there, ignoring his frantic family, until—just in time—JoJo manages to seize a clump of grass and pull himself up to where the ground is level. And—just in time—Rick gets back over the fence.

That evening the video was beamed across North American television channels. The director of JGI saw it. He called Rick. "That was a very brave thing you did. What made you do it?"

"Well you see," replied Rick, "I happened to look into his eyes, and it was like looking into the eyes of a man. And the message was 'Won't *anybody* help me?'"

I have seen that appeal for help in the eyes of so many suffering creatures. An orphan chimp tied up for sale in an African market; an adult male looking out from his five-by-five-foot sterile cell in a medical research laboratory; a dog, emaciated and starving, abandoned by her owner on the beach in Dar es Salaam; an elephant chained to a cement floor by one front and one hind foot. I've seen it in the eyes of street children, and those who have seen their families killed in the "ethnic cleansing" in Burundi. All around us, all around the world, suffering individuals look toward us with a plea in their eyes, asking us for help.

And if we dare to look into those eyes, then we shall feel their suffering in our hearts. More and more people have seen that appeal and felt it in their hearts. All around the world there is an awakening of understanding and compassion, an understanding that reaches out to help the suffering animals in their vanishing homelands. That embraces hungry, sick, and desperate human beings, people who are starving and dying while the fortunate among us—anyone who can afford to buy this book—have so much more than we need. And if, one by one, we help them, the hurting animals, the desperate humans, then together we shall alleviate so much of the hunger, fear, and pain in the world. Together we can bring change to the world, gradually replacing fear and hatred with compassion and love. Love for all living beings.

CODA

AFTER ALL IS SAID AND DONE, SILENCE IS BETRAYAL

It's one o'clock in the afternoon in Boulder, and nine at night in Bournemouth, where Jane has a few days of "rest" at home. "Rest" means working only about eighteen hours a day. I call and ask her how she is, and she tells me she is tired. Jane was invited to represent the environment at the then upcoming Winter Olympics in Salt Lake City, Utah, and she had to decline because she learned that rodeo was going to be performed there—as an example of American culture, as a way to celebrate "mutual understanding, tolerance, and brotherhood throughout the world." Forty cowboys from the United States and forty from Canada were slated to compete for $140,000 in prize monies, whereas the real Olympic athletes do not get paid for partaking in the games.

Tolerance and understanding for whom? Rodeo can be an incredibly inhumane activity—horses are stimulated with electric prods and have their tails twisted and testicles cinched to make them buck—and in some rodeos children partake in "mutton busting" to prepare them for their adult days in rodeo. Animals also suffer broken bones, lameness, and abrasions, and are choked by ropes tightened around their neck. In addition, as if more discouraging news were needed, despite the fact that a coyote named Copper was one of the game's mascots, coyotes are still being killed for bounty in Utah, using horrific methods including aerial shooting, poisoning, trapping, and snaring. They were to be spared only temporarily because, for security reasons, there was a no-fly zone over some of their habitat during the Olympics.

The other Olympic mascots, a snowshoe hare named Powder and a black bear named Coal, also represent animals persecuted in Utah. Since I've been studying coyotes for more than thirty years, I was able to help Jane along. At five the next morning I faxed Jane some information about the Olympic rodeo, including a wonderful letter written by a former gold-medal winner in skating, Scott Hamilton (which we mention in the First Trust), along with an essay I'd recently written about coyotes, those cunning, tireless tricksters who have been able to survive unrelenting human onslaughts. But our appeals to stop the rodeo to the Olympic Committee fell on deaf ears. Unfortunately, other forms of "rodeo" exist. At a meeting of the Performing Animal Welfare Society (PAWS) in May 2002, I saw a video of a man surfing on a captive dolphin's back—how undignified for both the dolphin and the man.

Some people have asked me how this book got written. I began writing the first draft on April 20, 2001, my mother's eighty-first birthday. I chose that day in honor of my extraordinary mother, Beatrice. Writing the manuscript was much fun, but also a lot of hard work. Finding good stories and blending in relevant scientific data was a challenge. At the same time, Jane was taking up the Trusts, adding her deep concern and wisdom from her experience and knowledge of animal behavior. Needless to say, there were countless transatlantic (and transworld) telephone calls and faxes. Often I would call Jane when she was surrounded (but obviously not defeated) by mounds of paper that were being cut and pasted into the manuscript (which I would later have to type). It was uncanny how often Jane and I were working on the same Trust and having the same thoughts! On two occasions I remember faxes containing the same material were simultaneously sent. Once, as we were talking about the number of accredited zoos in Europe, I received an e-mail from our Belgian friend Koen Margodt, from whom I had requested this information, giving me the details. This was one of a number of synchronicities that made for light moments along the way. But in retrospect these coincidences are not all *that surprising, for Jane and I share a deep commitment to make Earth a better place for all.*

It is common for Jane and me to communicate at odd hours—early in the morning for me and late at night for her. At four in the morning on February 1 of this year I called Jane to tell her about what was happening with IBP (see Trust Two) and the misuse of moneys that Senator Robert Byrd had allocated for enforcement of the Humane Slaughter Act. Gail Eisnitz, author of Slaughterhouse, *the widely acclaimed exposé on the meat industry, along with Kathy Liss and Chris Heyde at the Animal Welfare Institute and other devoted people at the Humane Farming Association, needed help, and Jane and I were able to lend a hand. In the midst of all that was happening, Jane called Senator Byrd and wrote him a letter supporting his efforts to bring about major reform in the treatment of animals condemned to slaughterhouses. In February 2002 Senator Byrd went to a meeting of the Agriculture Appropriations Subcommittee and asked Agriculture Secretary Ann Veneman questions about the inhumane slaughter of millions of animals. She was unable to answer such simple questions as those dealing with how many violations of the Humane Slaughter Act had occurred, how many slaughter plants exist in the United States, and how the $3 million that Senator Byrd allocated to improve the well-being of slaughterhouse animals was actually spent. Senator Byrd told the committee: "If I am still living a year from now and I'll be back, be prepared to answer these questions. These animals cannot speak for themselves."*

It also turns out, I learned, that slaughterhouses in the United States earn about 30 percent of their profit by selling animal hides to leather tanneries, and that the National Football League (NFL) in the United States uses more the twelve thousand cows for footballs. People living near tanneries suffer higher rates of leukemia and testicular cancer than those who do not live near these facilities.

I live in the mountains outside of Boulder. I love where I live because I am a "biophiliac"—a lover of all nature—at heart. A few years ago I had a window installed that allows me to look at a magnificent ponderosa pine tree. When I asked my friend to do the carpentry, he was incredulous—"You'll just see the darned tree," he told me. As if I didn't know! "I know," I told him, "I love trees! I can see mountains from other windows, but seeing and feeling the presence of this tree makes me feel good—makes me smile—makes me appreciate all of nature." Often I just sit and stare at "tree" and wonder what it is feeling.

Trees are wonderful beings, and they provide all sorts of comfort for many animals. They are sought as homes and refuges, but as we wrote above, refuges are not always safe havens. In our Second Trust, Jane tells of the horror of seeing a treed cougar shot in Wyoming. Because I've also chosen to live among cougars, I've had a number of close encounters with them. Once, on a very dark and starry night, I got out of my car to say hello to my neighbor's German shepherd, Lolo, only to realize when I heard Lolo barking behind me that I was saying hello to a large male cougar who had just killed a red fox below my house! A few years ago I almost stepped on a cougar as I walked backward down my road, telling my neighbor that there was a cougar in the neighborhood and that he should watch his kids and his dog! Since then, I've merely changed my ways so that I can coexist with the magnificent beasts—cougars, black bears, coyotes, red foxes, deer, and many species of birds and insects—who have allowed me to move into their neighborhood. I now hike with my canine companion, Jethro, carrying bear spray and a flashlight; often I put a flashing red light and a bell on Jethro's collar—simple adjustments indeed.

Let us not forget that in most instances we have intruded on other animals—they are not the intruders. And many animals suffer each and every day because of the messes we make. In a sense Earth is part of an uncontrolled "experiment" in which each of us plays a part. In May 2001, Jane and I were sitting at an outdoor cafe in Paris, France (at a meeting of the group Science and the Spiritual Quest II), when we saw a pigeon hobbling along, one of her legs wrapped in a piece of wire. We tried to catch her, but she evaded us. There was nothing we could do but lament this poor bird's plight. Birds also get their heads stuck in plastic bottle carriers, and other animals often suffer from trash left behind at campsites and in oceans, rivers, and lakes. Not only are humans *all over the place, but so is our trash. Many of us, but surely not all of us, simply have too much "stuff."*

We are not alone on this planet, although we frequently behave as if we were. Our big old slowly evolving brains that are confronted by new and rapidly evolving sociocultural milieus not only keep us somewhat in contact with nature but also remove us from nature, and this alienation results in our wanton abuse of the Earth. We are continually faced with making difficult and oftentimes agonizing choices that have short-term and long-term consequences. And we must look at the overall effects of our activities. Our influence is not always evident even in the short-term. For example, global warming appears to be having an impact on the number of primates living in Ethiopia. A rise in temperature leads to less grass and fewer crops on which gelada baboons are able to graze, and it is feared that their numbers will fall because of this. In addition to causing decreased numbers of gelada baboons, drought also reduces play among juveniles, and the lack of play in youngsters may have large effects on the social behavior and social organizations of adults.

In addition to a wide variety of mammals, birds also suffer from climatic warming and the recession northward of the polar ice pack. Long-term studies are needed to show the effects of climate change. George Divoky's three decades of research on Cooper Island in the Arctic has shown that pigeon-sized seabirds, black guillemots, have suffered because as the ice edge moves offshore and retreats northward, individuals are unable to reach the ice edge and consequently die. The extent of ice in the Arctic Ocean decreased 3 percent per decade between 1978 and 1996, and the sea ice in the Nordic seas has decreased about 30 percent during the past 130 years. It has been predicted that summer ice in the Arctic Ocean might shrink as much as 60 percent as carbon dioxide doubles.

The abuse and killing of animals continues, but we must make our objections known. Given what some people choose to do to animals, I often wish they were not the marvelous and magical beings they are. But the fact is that many animals do indeed suffer brutal exploitation and intense pain, and we must change our ways—now. We need animals and love animals because they are feeling beings, not because they are unfeeling "things." Of course, even animals who might not be able to experience pain and suffering deserve our respect and consideration so that their lives are not compromised by our self-centered activities. We do not have to be afraid of being sentimentalists or apologize for our idealism. Perhaps it seems to no avail at the moment, but as more and more people object to animal abuse and the exploitation of our environment, our children all will reap the benefits. We must remain hopeful that a universal ethic of courage, caring, sharing, respect, radical compassion, and love will make a difference even if we do not see the positive results of our efforts. Frustrations and personal insults need to be pushed aside, for tangible rewards do not frequently immediately follow activism for animals and the Earth.

We must also deeply believe that the voice and the actions of every individual also make a difference, for they do. Martin Luther King, Jr., once said: "A time comes when silence is betrayal." He was right—silence and indifference can be deadly for our animal friends and for the Earth.

There is an old saying: "After all is said and done, much more's been said than done." Although this is so for our interactions with animals, the human community, and the Earth and is among the many reasons Jane and I wrote this book, we have indeed made much progress in making this a better world, a more compassionate world in which caring and sharing abound. By "minding animals" and "minding the Earth" numerous animals, people, and habitats are far better off than they would have been in the absence of an ethic blending together respect, caring, compassion, humility, grace, and love. Caring about some being or some thing, any being or any thing, can spill over into caring for everybody and everything. If we focus on the awe and mystery of other animals and the Earth, perhaps we will be less likely to destroy them.

Allowing ourselves to sense the presence of other animals, to feel their residence in our hearts, brings much joy and peace and can foster spiritual development and a sense of unity. And this happiness, this sense of bliss, allows for Earth, bodies of water, air, animals, and people to be blended into a seamless tapestry, a warm blanket of caring and compassion, in which every single individual counts and every single individual makes a difference. The *interconnectedness of individuals in a community means that what one does affects all—what happens in New York influences what happens across the world in Beijing and other distant locales.*

Recently I read a report that emphasized just how interconnected we all are. Scientists from the U.S. Geological Survey discovered that dust from the Sahara Desert is blown across the Atlantic Ocean and found in the Caribbean and in the United States. This dust carries with it tiny microbes that survive the five- to seven-day journey. Bacteria, fungi, and viruses actually hitch a ride across the ocean and find their way to locations thousands of miles away. As a result, there is a heightened risk of respiratory diseases. One study found that on a dusty day there was an average of 158 bacteria and 213 viruses in a quart of air, whereas on a clear day there was an average of only 18 bacteria and 18 viruses in the same volume of air.

Back in 1963, President John F. Kennedy spoke the following words, which still apply today: "For in the final analysis, our most basic common link is that we all inhabit this small planet. We all breathe the same air. We all cherish our children's future. And we are all mortal." We are the community of Earth— we only have one Earth—and we need community now more than ever.

As I am writing this brief conclusion and getting weary—long days, horrible animal abuse here and there, incessant environmental damage due to human activities—I am rekindled when I receive an e-mail telling me of a team of more than two hundred Nahua Indian women in the Mexican state of Puebla who are involved in a wonderful environmental program that involves running a hotel and managing a health-food store, a greenhouse for medicinal plants, and a crafts store in which they produce recyclable paper and biodegradable cleaning products. And the good news continues. In May 2002 Germany became the first European country to vote to guarantee constitutional rights to animals. And Mexico became the world's largest national whale sanctuary when it signed an accord to protect whales in its waters, an area of about 1.1 million square miles. And in India, the newly founded National Institute of Animal Welfare now offers a B.A. degree in animal welfare. A Japanese fisherman recently refused to kill dolphins because "they cry when they are about to die . . .

I cannot kill something when tears are rolling down its cheeks." In surveys of more than 1,500 people in Europe and the United States, 70 to 90 percent of the respondents "recognize the right of nature to exist even if not useful to humans in any way." There is a new toy, called the Talk To Me Treatball, that enables people to record their voices and leave food for dogs and cats who are left alone for long periods of time. This simple toy relieves separation anxiety from which many household companions suffer when they are left alone at home. The mayor of Cincinnati, Ohio, recently turned over the key to his city to a cow who had escaped from a local slaughterhouse rather than return her to certain and painful death. Charles River Laboratories, a major supplier of millions of laboratory animals, has reduced the animal portion of their business from 80 to 40 percent since 1997. And in April 2002, Charles Schwab & Company, the world's largest online broker, severed its relationship with Huntington Life Sciences, a laboratory well-known for performing vivisection on animals. And at about the same time the prestigious scientific journal Nature called for more research on the nature of animal suffering and animal cognition. Although there is still much work to be done and caution needs to temper wishful thinking, this is a promising beginning to eliminating animal use, and pressure needs to be exerted to make sure this trend continues.

On a visit in April 2002 to Vancouver, Canada, to talk to members of the British Columbia Society for the Prevention of Cruelty to Animals (BCSPCA), incredible people who work tirelessly for all animals, about the importance of play behavior for the well-being of animals, I was stilled when I heard the story of Kevin Anderson, a twenty-nine-year-old who two months earlier had stopped his car while driving on a rainy night in Crescent Beach (British Columbia) to rescue a dog who was on the road. After getting out of his car and picking up the dog Kevin was struck by a car and killed—the dog lived. Kevin was posthumously given the Heroism Award by the BCSPCA. I had the honor of meeting Kevin's family, a bereaved group of wonderful human beings who are so proud of Kevin for the wonderful man he was and whose legacy lives on. A mark of courage and true heroism is how one acts when no one notices. Kevin was truly a courageous hero for whom being kind took precedence over everything else. I wish I had known Kevin. The world would be a much better place if there were more people like him who acted selflessly on behalf of animals.

And there is Zivvy Epstein, who loves rain forests and wants to spend his life trying to save them. Zivvy is part of the Roots & Shoots program that I helped to organize with Addie Rocchio and Stacey Selcho at the Collage Children's Museum in Boulder. At his seventh birthday party Zivvy collected $442 rather than gifts and gave this money to the Rainforest Alliance. And the students at the public schools of Toronto, Ohio, produced a wonderful book of poems, essays, and drawings as a tribute to September 11, 2001.

Kids are tireless crusaders for goodness and peace. And so are others who are often "written off," senior citizens and prisoners, people who also have hopes and dreams. I've worked at Golden West seniors' residence, where elders are eager to share their views about animals, people, and the environment. Many elders still have a spark of life that is contagious, and they get animated and feel empowered because others care about what they have to say and about what they do. Likewise, prisoners. My work at the Boulder County Jail inspires and energizes me. I feel I am a better person because of my interactions with elders and with inmates; I am sure I learn as much from them as they learn from me.

LET US BE ETHICISTS, NOT OSTRICHES

As a child of the sixties, I am a dreamer with few bounds. I ask the people with whom I interact to imagine that they carry a suitcase of courage, compassion, and hope and that because they receive what they give, the supply of courage, compassion, and hope will never be exhausted. It is easy to have one's spirit and soul weathered by the "bad" things that happen around us. It seems as if we are addicted to the destruction of the very animals and landscapes we love. But many, many good things are happening each and every day all over the world that can kindle our spirit and impel us to act.

Jane and I hope that we have inspired you to act—to do something, anything—to make this a better world. As Margaret Mead noted: "Never doubt that a small group of thoughtful, committed citizens can change the world. Indeed, it is the only thing that ever has." It is important to focus your efforts and not be deflected by those who oppose you. It's a waste of time and energy to "fight" your opponents who only gain when you divert your attention to them rather than to the important issues at hand. It is never too late to do something. Even if you have only one minute, or ten seconds a day, you can make a difference. Talk to friends and families while taking a break, while taking a walk, while just "hanging out." Shut off lights, take shorter showers, walk rather than drive, recycle, or say "hello" to a passerby. Write letters to local media. The small fraction that we each offer can contribute to larger solutions. Even a tiny ripple—a little agitation—can spread widely and rapidly. Even if you have time to help only one individual, you can make a difference. It is thought that North Atlantic right whales might survive if only one or a few females are spared each year—one whale counts.

And be resourceful. Use a blend of scientific data, common sense, and anecdotes to assess information and to make choices about what to do in a given situation. In this high-tech world there are innumerable sources available almost instantaneously. Jane told me about an introduced plant that was overtaking Buffalo, New York, but could not remember its name. I called a bed and breakfast establishment in Buffalo, figuring that those in the tourism business might know about some of the local ecological problems, and they did! I learned that it was the purple loosestrife (see our Fifth Trust), and that many people are indeed concerned about this nasty invasive plant that is rampantly taking over native plants in many regions. My sense of despair was replaced with a sense of hope. I think some people get discouraged and lose hope because they do not have the facts, and after talking to some of my colleagues about this and other problems that I thought were thoroughly unsolvable, I discovered that they could be dealt with if people knew what was really happening. And when we are unsure, I favor erring on the side of animals and nature.

The innumerable problems that currently exist will not disappear if we bury our heads in the sand like ostriches. The situation will only get worse. Achieving win-win situations for all humans and animals (and other nature) in the labyrinth of conflict and confusion will be very difficult, but we should never stop trying. If we fail to do so, I fear that we, our children and theirs, other animals, and nature as a whole will lose, and that much of the spark, spirit, and soul that keeps us going in our attempts to make this a better world will be extinguished. Fortunately, it is my impression, and others agree, that more students and people in general are now interested in ethical issues than had previously been the case, and that there is a progressive trend toward caring more and doing more, not less, for animals, people, and the Earth. I am amazed and thrilled that a search on the Worldwide Web for "animal protection," "conservation," or "biodiversity" produces tens of thousands of "hits."

Do we really want to live in harmony with nature? Are we truly the people we think we are? These are simple yet extremely challenging questions. If we answer yes to either or both, which not only is politically correct but also ethically and ecologically correct, we are compelled to move forward with grace, humility, respect, compassion, and love. We will need to replace "mindlessness" with "mindfulness" about our interactions with animals and the Earth. Nothing will be lost and much will be gained. We can never be too generous or too kind.

Surely, we will come to feel better about ourselves if we know deep in our hearts that we did the best we could and took into account the well-being of the magnificent animals with whom we share Earth, the awesome and magical beings who selflessly make our lives richer, more challenging, and more enjoyable than they would be in the animals' absence. Doesn't it feel good to know that there are animals "out there" whom we have helped even if we cannot see them? Doesn't it feel good to know that we did something to help the Earth even if we do not see the fruits of our labor? As we attempt to restore nature, we can restore ourselves, our psyches that have been fragmented because of our alienation from animals and other nature. We often turn toward nature when we're feeling down. We need animals, nature, and wildness.

We need to wage peace. Caring for others also brings much peace, and this peace can spread rapidly and widely among all peoples. Peace and reconciliation are at the top of the agendas of many world leaders. In April 2002, Jane was named a "Messenger for Peace" by Kofi Annan, the Secretary General of the United Nations. I was invited to give a lecture at a meeting concerned with "The Path to Nature's Wisdom," convened by His Holiness the Dalai Lama as part of his Kalachakra for World Peace 2002. Peace and calm (and personal happiness, as stressed by His Holiness the Dalai Lama) are needed to bring individuals into a community in which differences are pushed aside for the common goal of making this a better world for all beings.

Let us never avert our eyes from the gaze of the animals who need us and whom we need as much or more. Life without our animal friends would be lonely and miserable. In the grand scheme of things, individuals receive what they give. If love is poured out in abundance, then it will be returned in abundance. There is no need to fear depleting the potent and self-reinforcing feeling of love that continuously can serve as a powerful stimulant for generating compassion, respect, and more love for all life. Each and every individual plays an essential role and that individual's spirit and love are intertwined with the spirit and love of others. These emergent interrelationships transcending individuals foster a sense of oneness. These interrelationships can work in harmony to make this a better and more compassionate world for all beings. We must stroll with our kin and not leave them in the wake of our tumultuous, rampant, self-serving destruction.

By minding animals we mind ourselves. Personal transformations are needed and will serve us well. We owe it to future generations to transcend the present, to share dreams for a better world, to step lightly, to move cautiously with restraint. We destroy one another when we destroy the Earth. We all can be dreamers and doers. We owe it to ourselves and to other animals, to whom we can, unfortunately, do whatever we choose. We owe it to ourselves to keep in mind the power of love. As big-brained, omnipresent, powerful, and supposedly omniscient mammals, we are the most powerful beings on Earth. We really are that *powerful*, and with that might are inextricably tied innumerable staggering responsibilities to be ethical human beings. We can be no less.

MARC BEKOFF

ACKNOWLEDGMENTS

Many people helped us find material and also discussed some of the issues on which we write. It is impossible to remember all of them, but those who provided information that was extremely difficult to locate or who went beyond the call of duty include Chris Heyde, Koen Margodt, Tony Smith, Marty Becker, David Baron, Jim Willis (who with his wife, Nicole Valentin-Willis, runs the Tiergarten Sanctuary Trust), Matt Rossell, Dale Peterson, Joan Kleypas, Aimee Morgana, Brenda Peterson, Tom Mangelsen, Cara Blessley Lowe, Tom Ranker, Gary McAvoy, Jessica Almy, and Gail Eisnitz. We also thank our agent, Jonathan Lazear, and his delightful co-workers, Christi Cardenas, Julie Mayo, and Tanya Cromley, for their help in launching this project and keeping it afloat. Julie Mayo miraculously typed the penultimate draft from a collage of pasted notes. Mie Horiuchi, Communications Coordinator at the Jane Goodall Institute, was always there to provide logistical support, and Liz Perle, our editor at Harper San Francisco, was a pleasure to work with. Anne Connolly and Terri Leonard, also at Harper San Francisco, were awesome intermediaries who were always available to help. And of course, Mary Lewis was *always* there, fetching Jane for phone calls, letting Marc know where Jane was and when she would be available, telling Marc when more water heating coils were needed, sending e-mails with pertinent information, and providing innumerable and much-appreciated light moments.

SOURCES

Here are some suggested readings and Web sites that we used as resources for our book. This list is not meant to be exhaustive, and numerous other essays and Web sites can be found in the reference sections of these sources.

BOOKS AND ARTICLES

Abram, D. 1996. *The Spell of the Sensuous: Perception and Language in a More-Than-Human World*. New York: Pantheon Books.

Allen, C., and M. Bekoff. 1997. *Species of Mind: The Philosophy and Biology of Cognitive Ethology*. Cambridge, MA: MIT Press.

Ammann, K. 2001. "Bushmeat Hunting and the Great Apes." In B. Beck et al., eds., *Great Apes and Humans: The Ethics of Coexistence*. Washington, D.C.: Smithsonian Institution Press, pp. 71–85.

Animals Agenda Directory of Organizations. 2002. New York: Lantern Books.

Balcombe, J. 1999. *The Use of Animals in Education: Problems, Alternatives, and Recommendations*. Washington, D.C.: Humane Society of the United States.

Balls, M., A.-M. van Zeller, and M. E. Halder, eds. 2000. *Progress in the Reduction, Refinement and Replacement of Animal Experimentation*. The Netherlands: Elsevier.

Barrett, L., R. I. M. Dunbar, and P. Dunbar. 1992. "Environmental Influence on Play Behaviour in Immature Gelada Baboons," *Animal Behaviour* 44: 111–15.

Beck, A., and A. Katcher. 1996. *Between Pets and People: The Importance of Animal Companionship*. West Lafayette, IN: Purdue University Press.

Becker, M. 2002. *The Healing Power of Pets*. New York: Hyperion.

Bekoff, M., ed. 1998. *Encyclopedia of Animal Rights and Animal Welfare*. Westport, CT: Greenwood.

———. 2000. *Strolling with Our Kin: Speaking for and Respecting Voiceless Animals*. New York: AAVS/Lantern Books. (Available in Italian, German, Turkish, Chinese, and Spanish.)

———, ed. 2000. *The Smile of a Dolphin: Remarkable Accounts of Animal Emotions*. New York: Discovery Books.

———. 2000. "Field Studies and Animal Models: The Possibility of Misleading Inferences." In M. Balls, A.-M. van Zeller, and M. E. Halder, eds., *Progress in the Reduction, Refinement and Replacement of Animal Experimentation*. The Netherlands: Elsevier, pp. 1553–59.

———. 2002. *Minding Animals: Awareness, Emotions, and Heart*. New York: Oxford University Press.

Berger, J., A. Hoylman, and W. Weber. 2001. "Perturbation of Vast Ecosystems in the Absence of Adequate Science: Alaska's Arctic Refuge," *Conservation Biology* 15: 539–41.

Berry, T. 1999. *The Great Work*. New York: Bell Tower.

Boone, J. A. 1976. *Kinship with All Life*. San Francisco: Harper & Row Publishers.

Brawn, J. D. et al. 2001. "The Role of Disturbance in the Ecology and Conservation of Birds." *Annual Reviews of Ecology and Systematics* 32: 251–76.

Brown, D. A. 2002. *American Heat: Ethical Problems with the United States' Response to Global Warming*. Lanham, MD: Rowman & Littlefield.

Brown, L., M. Renner, and B. Halwell. 2000. *Vital Signs: The Environmental Trends That Are Shaping Our Future*. New York: Norton.

Burgess-Jackson, K. 1998. "Doing Right by Our Animal Companions," *Journal of Ethics* 2: 159–85.

Campbell, T. C., and C. J. Chen. 1994. "Diet and Chronic Degenerative Diseases: Perspectives from China," *American Journal of Clinical Nutrition* 59: 1153–61.

Cantalupo, C., and W. D. Hopkins. 2001. "Asymmetric Broca's Area in Great Apes," *Nature* 414: 505.

Caro, T., ed. 1998. *Behavioral Ecology and Conservation Biology.* New York: Oxford University Press.

Cavalieri, P. 2001. *The Animal Question.* New York: Oxford University Press.

Cavalieri, P., and P. Singer, eds. 1993. *The Great Ape Project: Equality Beyond Humanity.* London: Fourth Estate.

Cheney, D. L., and R. M. Seyfarth. 1990. *How Monkeys See the World: Inside the Mind of Another Species.* Chicago: University of Chicago Press.

Constable, J. L., M. V. Ashley, J. Goodall, and A. E. Pusey. 2001. "Noninvasive Paternity Assignment in Gombe Chimpanzees," *Molecular Ecology* 10: 1279–1300.

Crooks, K., and M. E. Soulé. 1999. "Mesopredator Release and Avifaunal Extinctions in a Fragmented System," *Nature* 400: 563–66.

Curtis, P. 2002. *City Dog: Choosing and Living Well with a Dog in Town.* New York: Lantern Books.

D'Agnese, J. 2002. "An Embarrassment of Chimpanzees." *Discover,* May.

Dalai Lama, His Holiness the. 1999. *The Path to Tranquillity: Daily Wisdom.* New York: Viking Arkana.

Darwin, C. 1872/1998. *The Expression of the Emotions in Man and Animals,* 3d ed. With an introduction, afterword, and commentaries by Paul Ekman. New York: Oxford University Press.

Davis, K. 1996. *Poisoned Chickens, Poisoned Eggs: An Inside Look at the Modern Poultry Industry.* Summertown, TN: Book Publishing Company.

de Waal, F. 2001. *The Ape and the Sushi Master.* New York: Basic Books.

Douglas-Hamilton, I., and O. Douglas-Hamilton. 1975. *Among the Elephants.* New York: Viking.

Ehrlich, P. R. 1997. *A World of Wounds: Ecologists and the Human Dilemma.* Oldendorf/Luhe, Germany: Ecology Institute.

Eisnitz, G. A. 1997. *Slaughterhouse: The Shocking Story of Greed, Neglect, and Inhumane Treatment Inside the U. S. Meat Industry.* Buffalo, NY: Prometheus Books.

Fano, A. 1997. *Lethal Laws: Animal Testing, Human Health and Environmental Policy.* London: Zed Books.

Fossey, D. 2000. *Gorillas in the Mist.* Boston: Houghton Mifflin, Mariner Books.

Fouts, R., with S. Mills. 1997. *Next of Kin: What Chimpanzees Have Taught Me About Who We Are.* New York: Morrow.

Fox, M. A. 1999. *Deep Vegetarianism.* Philadelphia: Temple University Press.

Fox, M. W. 1997. *Eating with Conscience: The Bioethics of Good.* Troutdale, OR: NewSage Press.

———. 1999. *Beyond Evolution: The Genetically Altered Future of Plants, Animals, the Earth . . . and Humans.* New York: Lyons Press.

Francione, G. L. 1995. *Animals, Property, and the Law.* Philadelphia: Temple University Press.

———. 2000. *Introduction to Animal Rights: Your Child or the Dog?* Philadelphia: Temple University Press.

Francione, G. L., and A. E. Charlton. 1992. *Vivisection and Dissection in the Classroom: A Guide to Conscientious Objection.* Jenkintown, PA: The American Anti-Vivisection Society.

Frey, D. 2002. "George Divoky's Planet," *New York Times Magazine,* January 6.

Galdikas, B. M. F. 1996. *Reflections of Eden: My Years with the Orangutans of Borneo.* Boston: Little, Brown.

Glassner, P. M., ed. 2001. *Cinderella Dogs: Real-Life Fairy Tail Adoptions from the San Francisco SPCA.* San Francisco: Kinship Communications.

Glenz, C., A. Massolo, D. Kuonen, and R. Schlaepfer. 2001. "A Wolf Habitat Suitability Prediction Study in Valais (Switzerland)," *Landscape and Urban Planning* 55: 55–65.

Goodall, J. 1986. *The Chimpanzees of Gombe.* Cambridge, MA: Harvard University Press.

———. 1990. *Through a Window: My Thirty Years with the Chimpanzees of Gombe.* Boston: Houghton Mifflin.

———. 1999. *Reason for Hope: A Spiritual Journey.* New York: Warner Books.

Greek, R., and J. Greek. 2000. *Sacred Cows and Golden Geese*. New York: Continuum.

Green, A. 1999. *Animal Underworld: Inside America's Market for Rare and Exotic Species*. New York: Public Affairs.

Griffin, D. R. 2001. *Animal Minds: Beyond Cognition to Consciousness*. Chicago: University of Chicago Press.

Grimaldi, J. V. 2002. National Zoo cites privacy concerns in its refusal to release animals' medical records. *Washington Post*, May 6, pg. E12.

Guillermo, K. S. 1993. *Monkey Business: The Disturbing Case That Launched the Animal Rights Movement*. Washington, D.C.: National Press Books.

Hancocks, D. 2001. *A Different Nature: The Paradoxical World of Zoos and Their Uncertain Future*. Berkeley: University of California Press.

Hauser, M. 2000. *Wild Minds: What Animals Really Think*. New York: Holt.

Heseltine, A. 2002. "The Blood of Dolphins," *Earth Island Journal* 17: 24.

Hill, Julia Butterfly. 2000. *The Legacy of Luna*. San Francisco: HarperSanFrancisco.

Huffman, M. A. 2001. "Self-Medicative Behavior in the African Great Apes: An Evolutionary Perspective into the Origins of Human Traditional Medicine," *BioScience* 51: 651–61.

Jensen, D. 2000. *A Language Older Than Words*. New York: Context Books.

Kareiva, P. 2001. "When One Whale Matters," *Nature* 414: 493–94.

Kennedy, J. F. 1963. Speech at American University. June 10.

Key, M. H. 2001. *What Animals Teach Us*. Roseville, CA: Prima Publishing.

Kincade, A. 2001. *Straight from the Horse's Mouth*. New York: Crown. (This book provides a very useful list of manufacturers of cruelty-free products.)

Kistler, J., ed. 2000. *Animal Rights: Subject Guide and Bibliography with Internet Sites*. Westport, CT: Greenwood.

Knight, A. 2002. *Learning Without Killing: A Guide to Conscientious Objection*. www.interniche.org.

Koren, C., et al. 2002. "A Novel Method Using Hair for Determining Hormonal Levels in Wildlife," *Animal Behaviour* 63: 403–6.

Krause, B. 2002. "The Loss of Natural Soundscapes." *Earth Island Journal*, Spring.

Laszlow, E. 2001. *Macroshift: Navigating the Transformation to a Sustainable World.* San Francisco: Berrett-Koehler.

Linnell, J. D. C. et al. 2001. "Predators and People: Conservation of Large Carnivores is Possible at High Human Densities if Management Policy is Favorable." *Animal Conservation* 4: 345–49.

Linzey, A. 1976. *Animal Rights.* London: SCM Press.

Mangelsen, T. D., story by C. S. Blessley. 1999. *Spirit of the Rockies: The Mountain Lions of Jackson Hole.* Omaha, NE: Images of Nature.

Lyman, H. F. 1998. *Mad Cowboy: Plain Truth from the Cattle Rancher Who Won't Eat Meat.* New York: Scribners.

Margodt, K. 2000. *The Welfare Ark: Suggestions for a Renewed Policy in Zoos.* Brussels: VUB University Press.

Mason, J. 1993. *An Unnatural Order: Uncovering the Roots of Our Domination of Nature and Each Other.* New York: Simon & Schuster.

Masson, J., and S. McCarthy. 1995. *When Elephants Weep: The Emotional Lives of Animals.* New York: Delacorte.

Matsuzawa, T., ed. 2001. *Primate Origins of Human Cognition and Behavior.* New York: Springer.

McElroy, S. C. 1995. *Animals As Teachers and Healers.* Troutdale, OR: NewSage Press.

McKinney, M. L. 2001. "The Role of Human Population Size in Raising Bird and Mammal Threat Among Nations," *Animal Conservation* 4: 45–57.

Mead, M. The source of her quotation on activism remains a mystery; see www.mead2001.org/faq_page.htm#quote.

Midgley, M. 1983. *Animals and Why They Matter.* Athens: University of Georgia Press.

Montgomery, S. 1991. *Walking with the Great Apes: Jane Goodall, Dian Fossey, and Biruté Galdikas.* Boston: Houghton Mifflin.

Moss, C. 2000. *Elephant Memories: Thirteen Years in the Life of an Elephant Family.* Chicago: University of Chicago Press.

Nature. 2002. "Rights, Wrongs, and Ignorance," 416: 351.

Nilsson, G. 1981. *The Bird Business: A Study of the Commercial Cage Bird Trade.* Washington, D.C.: Animal Welfare Institute.

Noddings, N. 2002. *Starting at Home: Caring and Social Policy*. Berkeley: University of California Press.

Norris, S. 2002. "Creatures of Culture? Making the Case for Cultural Systems in Whales and Dolphins," *BioScience* 52 (1): 9–14.

Paine, R. T., and D. E. Schindler. 2002. "Ecological Pork: Novel Resources and the Trophic Reorganization of an Ecosystem," *Proceedings of the National Academy of Sciences* 99: 554–55.

Patterson, C. 2002. *Eternal Treblinka: Our Treatment of Animals and the Holocaust*. New York: Lantern Books.

Peterson, B. 2000. *Build Me an Ark*. New York: Norton.

Peterson, D. 1989. *The Deluge and the Ark: A Journey into Primate Worlds*. Boston: Houghton Mifflin.

Peterson, D., and J. Goodall. 1993. *Visions of Caliban: On Chimpanzees and People*. Boston: Houghton Mifflin.

Plous, S., and H. Herzog. 2000. "Polls Show That Researchers Favor Lab Animal Protection." *Science* 209: 711.

———. 2001. "Reliability of Protocal Reviews for Animal Research." *Science* 293: 608–9.

Pollan, M. 2002. "Power Steer." *New York Times Magazine*, March 31.

Poole, J. 1996. *Coming of Age with Elephants*. New York: Hyperion Press.

———. 1998. "An Exploration of a Commonality Between Ourselves and Elephants," *Etica & Animali* 9/98: 85–110.

Posey, D. A., ed. 1999. *Cultural and Spiritual Values of Biodiversity*. Nairobi, Kenya: United Nations Environment Programme.

Rachels, J. 1990. *Created from Animals: The Moral Implications of Darwinism*. New York: Oxford University Press. (The quote about monkeys subjected to lethal doses of radiation in the Third Trust is from p. 132.)

Randour, M. L. 2000. *Animal Grace: Entering a Spiritual Relationship with Our Fellow Creatures*. Novato, CA: New World Library.

Regan, T. 1983. *The Case for Animal Rights*. Berkeley: University of California Press.

Rendell, L., and H. Whitehead. 2001. "Culture in Whales and Dolphins," *Behavioral and Brain Sciences* 24: 309–82.

Rifkin, J. 1992. *Beyond Beef: The Rise and Fall of the Cattle Culture*. New York: E. P. Dutton.

Rivera, M. 2000. *Hospice Hounds*. New York: Lantern Books.

Riyan, J. G., et al. 2001. "The New Biophilia: An Exploration of Visions of Nature in Western Countries," *Environmental Conservation* 28: 1–11.

Roberts, C. M. 2002. "Deep Impact: The Rising Toll of Fishing in the Deep Sea," *Trends in Ecology and Evolution* 17 (5): 242–45.

Roemer, G., C. J. Donlan, and F. Courchamp. 2002. "Golden Eagles, Feral Pigs, and Insular Carnivores. How Exotic Species Turn Native Predators into Prey," *Proceedings of the National Academy of Sciences* 99: 791–96.

Rollin, B. E. 1989. *The Unheeded Cry: Animal Consciousness, Animal Pain and Science*. New York: Oxford University Press. Reprint 1998, Iowa State University Press.

Russell, E. 2001. *War and Nature: Fighting Humans and Insects with Chemicals from World War I to Silent Spring*. New York: Cambridge University Press.

Russell, W. M. S., and R. L. Burch. 1959/1992. *The Principles of Humane Experimental Technique*. Wheathampstead, England: UFAW.

Ryder, R. D. 1989. *Animal Revolution: Changing Attitudes Towards Speciesism*. London: Blackwell.

Salem, D. J., and A. N. Rowan, eds. 2001. *The State of the Animals 2001*. Washington, D.C.: Humane Society of the United States.

Samsel, R. W., G. A. Schmidt, J. B. Hall, L. D. H. Wood, S. G. Shroff, and P. T. Schumaker. 1994. "Cardiovascular Physiology Teaching: Computer Simulations vs. Animal Demonstrations," *Advances in Physiology Education* 11: S36–46.

Schirf, D. L. 2000. "Mauritius kestral." www.mindspring.com/~slywy/mkestrel.html.

Schneider, S. H., and T. L. Roots, eds. 2002. *Wildlife Responses to Climate Change: North American Case Studies*. Washington, D.C.: Island Press.

Schoen, A. M. 2001. *Kindred Spirits: How the Remarkable Bond Between Humans and Animals Can Change the Way We Live*. New York: Broadway Books.

Seligman, M. E. P., S. F. Maier, and J. H. Geer. 1968. "Alleviation of Learned Helplessness in the Dog," *Journal of Abnormal Psychology* 73: 256–62. (The quote about learned helplessness in the Third Trust is from p. 256.)

Sewall, L. 1999. *Sight and Sensibility: The Ecopsychology of Perception.* New York: Tarcher/Putnam.

Shapiro, K. 1998. *Animal Models of Human Psychology: Critique of Science, Ethics and Policy.* Seattle: Hogrefe & Huber.

Sheldrake, R. 1999. *Dogs That Know When Their Owners Are Coming Home, and Other Unexplained Powers of Animals.* London: Hutchinson.

Sheldrake, R., and P. Smart. 2000. "Testing a Return-Anticipating Dog," *Anthrozoös* 13: 203–11.

Siddle, S., with D. Cress. 2002. *In My Family Tree: A Life with Chimpanzees.* New York: Grove Press.

Singer, P. 1990 *Animal Liberation,* 2d ed. New York: New York Review of Books.

——. 1998. *Ethics into Action: Henry Spira and the Animal Rights Movement.* Lanham, MD: Rowman & Littlefield.

Spinka, M., R. C. Newberry, and M. Bekoff. 2001. "Mammalian Play: Training for the Unexpected," *Quarterly Review of Biology* 76: 141–68.

Stein, T. 2002. "Bear's Death Places Zoo Under Scrutiny," *Denver Post,* February 15. www.denverpost.com/Stories/0,1002,53%257E403732,00.html.

Streever, B. 2002. "Science and Emotion, on Ice: The Role of Science on Alaska's North Pole," *BioScience* 52: 179–84.

Suzuki, D., and H. Dressel. 2002. *Good News for a Change: Hope for a Troubled Planet.* Toronto: Stoddart Publishing Company Ltd.

The National Anti-Vivisection Society. 2000. *Personal Care for People Who Care.* 10th ed. Chicago. (This is an excellent guide for choosing cruelty-free products.)

Tinbergen, N. 1951/1989. *The Study of Instinct.* New York: Oxford University Press.

Tobias, M., and K. Solisti, eds. 1998. *Kinship with the Animals.* Portland, OR: Beyond Words Publishers.

Vincent, A., and Y. Sadovy. 1998. "Reproductive Ecology in the Conservation and Management of Fishes." In Caro, T., ed. *Behavioral Ecology and Conservation Biology.* New York: Oxford University Press.

von Kriesler, K. 2001. *Beauty in the Beasts: True Stories of Animals Who Choose to Do Good.* New York: Tarcher/Putnam.

Waskon, R. M. 1994. "Best Management Practices for Manure Utilization," *Colorado State University Bulletin.*

Whiten, A., et al. "Cultures in Chimpanzees," *Nature* 399: 682–85.

Wielebnowski, N. 1998. "Contributions of Behavioral Studies to Captive Management and Breeding of Rare and Endangered Mammals." In Caro, T., ed. *Behavioral Ecology and Conservation Biology.* New York: Oxford University Press.

Wilkie, D. S. 2001. "Bushmeat Trade in the Congo Basin." In B. Beck et al., eds. *Great Apes and Humans: The Ethics of Coexistence.* Washington, D.C.: Smithsonian Institution Press, pp. 86–109.

Williams, Terry Tempest. 2001. *Red: Passion and Patience in the Desert.* New York: Pantheon Books.

Willis, J. 2002. *Pieces of My Heart.* www.infinitypublishing.com.

Wilson, E. O. 2002. *The Future of Life.* New York: Alfred A. Knopf.

Wise, S. 2000. *Rattling the Cage: Toward Legal Rights for Animals.* Cambridge, MA: Perseus Books.

Woodruff, D. S. 2001. "Declines in Biomes and Biotas and the Future of Evolution," *Proceedings of the National Acaedmy of Sciences* 98: 5471–76.

Yorio, P., et al. 2001. "Tourism and Recreation at Seabird Breeding Sites in Patagonia, Argentina: Current Concerns and Future Prospects." *Bird Conservation International* 11: 231–45.

VIDEO

"Natural Connections." Snohomish, WA: Howard Rosen Productions. (The data on the effects of the loss of tree cover around Puget Sound in the Fifth Trust are from this video.)

WEB SITES

The German edition of *Strolling with Our Kin* (*Das unnötige Leiden der Tiere*, published by Herder Spektrum) contains information about many European animal protection organizations. The World Animal Net Directory at www.worldanimalnet.org is the world's largest database of animal protection societies, with more than 13,000 listings and more than 6,000 links to Web sites.

Alive: www.jca.apc.org/alive/ (Japan)

Amboseli Elephant Research Project: www.elephanttrust.org

American Anti-Vivisection Society (AAVS): www.aavs.org

Animal Alliance of Canada: www.animalalliance.ca

AnimalKind: www.netcomuk.co.uk/~jcox/index.html (United Kingdom)

Animal Protection Institute (API): www.api4animals.org

Animal Responsibility Cyprus: www.geocities.com/RainForest/1862/

Animals' Agenda magazine, at which can be found much information about animal protection groups worldwide: www.animalsagenda.org

Animal Welfare Institute (AWI): www.awionline.org

Ark Trust: www.arktrust.org

Asian Animal Protection Network: www.aapn.org

Born Free Foundation: www.bornfree.org.UK

Bright Eyes Society: www.brighteyes.dk/ (Spain)

Center for Captive Chimpanzee Care and Kids for Chimps: www.savethechimps.org

Circle of Life Foundation: www.circleoflifefoundation.org

Coral reefs: *see*
www.aims.gov.au/pages/research/coral-
bleaching/scr2000/scr-00.html

Earth Charter: www.earthcharter.org/aboutus/

Earth Elders: www.earthelders.org

Elephant Sanctuary: www.elephants.com

Endangered Species Chocolate Company: www.chocolatebar.com

Ente Nazionale Protezione Animali: www.arpnet.it/~enpa/

Ethologists for the Ethical Treatment of Animals/Citizens for
Responsible Animal Behavior Studies (EETA/CRABS):
www.ethologicalethics.org

Fauna Foundation: www.faunafoundation.org

Friends of Animals (FOA): www.friendsofanimals.org

Genetically engineered drugs: *see* www.iddinternational.org;
www.members@tripod.com/diabetes_world and associated links

Genetically modified organisms: *see*
www.thecampaign.org; www.thecampaign.org/newscenter.htm and
associated links

Green Chimneys: www.pcnet.com/~gchimney/index.html

Humane Farming Association (HFA): www.hfa.org

Humane Society of the United States (HSUS): www.hsus.org

Humulin: *see* www.iddinternational.org;
www.members@tripod.com/diabetes_world and associated links

In Defense of Animals (IDA): www.idausa.org

International Fund for Animal Welfare: www.ifaw.org

International Primate Protection League (IPPL): www.ippl.org

Jellyfish: *see* www.oregonlive.com/news/99/12/st122304.html

Korea Animal Protection Society: www.koreananimals.org

Laboratory Primate Advocacy Group (LPAG): www.lpag.org

Maynard, Elly: *see* www.sirius.oneuk.com/Rome.html

McLibel case: *see*
www.mcspotlight.org/case/trial/verdict/index.html;
www.mcspotlight.org

Monkeys: *see* www.oregonlive.com/news/99/12/st122304.html

Monkey World Ape Rescue Centre:
www.sql.monkeyworld.org.uk/v2/topic.phtml?TopicID=
36&Template=standard

Nonanimal alternatives: *see,* among others, www.mindlab.msu.edu;
www.enviroweb.org/avar; www.aavs.org; www.hsus.org;
www.pcrm.org/issues/Animal_Experimentation_Issues/college_
alternatives.html; www.interniche.org

Orangutan Foundation International:
www.orangutan.org/home/home.php

People for the Ethical Treatment of Animals (PETA):
www.petaonline.org

Performing Animal Welfare Society (PAWS): www.pawsweb.org

Physicians Committee for Responsible Medicine (PCRM):
www.pcrm.org

Premarin: *see* www.members.aol.com/novenaann/premarin.htm;
www.ohahs.org/Premarin_Facts.html

Psychologists for the Ethical Treatment of Animals (PsyETA):
www.psyeta.org

Riverkeeper, Inc.: www.riverkeeper.org

Roots & Shoots: www.janegoodall.org;
www.janegoodall.org/rs/rs_history.html

Talk To Me Treatball: www.talktometreatball.com

Tiergarten Sanctuary:
www.geocities.com/pyrangel/tiergarten_sanctuary.htm

Universities Federation for Animal Welfare: www.ufaw.org.uk

Wildlife Trust of India: www.wildlifetrustofindia.org/index.html

ABOUT THE AUTHORS

JANE GOODALL is one of the world's leading conservationists. She and a group of very dedicated researchers have been studying chimpanzees for more than forty years at Gombe Stream in Tanzania. Jane has won numerous international awards for her tireless efforts to make this a better and more peaceful world for all beings, and she has published many books, including *The Chimpanzees of Gombe, In the Shadow of Man, Through a Window, Reason for Hope, Africa in My Blood*, and *Beyond Innocence*. Jane's work has been shown throughout the world on numerous documentaries and featured in innumerable publications. In 2002 Jane was named a "Messenger for Peace" by Kofi Annan, the secretary general of the United Nations.

MARC BEKOFF is a professor of biology at the University of Colorado. A former Guggenheim Fellow and Fellow of the Animal Behavior Society, he was presented with the society's Exemplar Award in 2000 for his major long-term contributions to the field of animal behavior. Marc is a regional coordinator for the Jane Goodall Institute's Roots & Shoots program and also a member, along with Jane, of Science and the Spiritual Quest II. Together they founded Ethologists for the Ethical Treatment of Animals/Citizens for Responsible Animal Behavior Studies (www.ethologicalethics.org). Marc has written or edited *Encyclopedia of Animal Rights and Animal Welfare, Strolling with Our Kin: Speaking for and Respecting Voiceless Animals, The Smile of a Dolphin: Remarkable Accounts of Animal Emotions*, and *Minding Animals: Awareness, Emotions, and Heart*, among other books. His work has been featured in *Time, Life, U.S. News and World Report, New York Times, New Scientist, BBC Wildlife, Orion, Scientific American*, on *48 Hours*, National Public Radio, BBC, Fox, Natur GEO, and in National Geographic Society and Discovery television specials. Marc's home page is http://literati.net/Bekoff.